Sustainable Development in Energy Systems

Brian Azzopardi
Editor

Sustainable Development in Energy Systems

Springer

Editor
Brian Azzopardi
MCAST Energy Research Group
Malta College of Arts, Science and Technology (MCAST)
Paola
Malta

ISBN 978-3-319-54806-7 ISBN 978-3-319-54808-1 (eBook)
DOI 10.1007/978-3-319-54808-1

Library of Congress Control Number: 2017935399

Printed on acid-free paper

This Springer imprint is published by Springer Nature
The registered company is Springer International Publishing AG
The registered company address is: Gewerbestrasse 11, 6330 Cham, Switzerland

Editorial Advisory Board

Preface

Recent large-scale penetration of distributed renewable energy sources (RES) within the electricity grids has resulted in the need for more supervisory control and operation. This is giving rise to different application scenarios such as microgrids, nanogrids or just islanding of electrical energy systems within a controlled environment. The main bottleneck is that energy systems simply cannot cope with demand–supply mismatch. Therefore, further integration of clean generation such as RES entails increasing levels of complexity not only on the electric power networks but also on the electricity market and regulations contributing to new business and management modelling, finance and investment schemes for sustainable energy. Hence, it is important that cost-effectiveness is also considered together with the security and high quality of supply for customers. This book includes different renewable and alternative energy systems integration solutions for mostly existing infrastructures.

Chapter "Overview on Microgrids: Technologies, Control and Communications" of this book introduces the concept of microgrids from three perspectives: (i) technologies, (ii) control and (iii) communications. Microgrids are an ideal platform to integrate renewable energy sources on the community level, allowing for prosumers full market participation. The aim of this chapter is to provide a brief overview on microgrids, including the state of art about the main motivation for the emergence of these grids. The chapter starts with a survey on some existing microgrids around the world, followed by a typical microgrid's architecture presentation. Then, in Chapter "Hybrid Nanogrid Systems for Future Small Communities", the principles of hybrid nanogrid systems are introduced for future small communities where the dynamic performances connected via ac and dc buses are explored in particular during islanding operation.

Various aspects of grid management such as interconnectivity to support neighbouring microgirds are discussed in Chapter "Interconnected Microgrid Systems for Remote Areas". A communication assisted multi-agent system (MAS) control scheme is proposed in Chapter "Distributed Agent-based

Coordinated Control for Microgrid Management" to establish the coordination between the power sharing and energy management through agent communication which further ensures the sustainable operation of microgrids by effectively controlling the inverters.

The sustainable development of regional energy systems is presented through a case study in Chapter "Sustainable Development of Energy Systems in the Baltic Region" for the Baltic States, Lithuania, Latvia and Estonia, which until recently operated in non-EU electricity frequency system, having comparatively low level of interconnections between the other countries in the region. One of the main aims of the EU internal market legal regulation, so-called the third energy market package (electricity and gas), is to develop well-functioning energy market, ensuring integrated net of infrastructure within entire EU territory. The idea foresees to create capable enough interconnections between all the EU member states therewith eliminating considerable dependence of particular countries on energy imports from non-EU countries, as well as ensure that the interconnected EU market operates according to harmonised EU principles and regulations. A closer integration of networks is particularly important to the states considered to be "energy islands", when electricity and gas networks are not connected with the other EU member states.

The most important integration bottlenecks for RES and its contribution to energy systems sustainable development are access to the market and the grid itself. A number of renewable energy sources and configurations are addressed in the next four chapters. Chapters "Active Distribution Networks Operation within a Distribution Market Environment" and "Critical Performance Evaluation of a Few Intelligent Controllers for Effective Power Control in Variable Speed Wind Generators" address this through wind energy integration, Chapter "Power Flow Constrained Short-Term Scheduling of CHP Units" considers combined heat and power units (CHPs) and Chapter "Optimal Utilization of Solar Energy Resources in Hungary" addresses this through solar resource analysis in Hungary. The methodologies vary from technical, semi-technical to techno-economic analysis such as optimal power flow in Chapters "Active Operation within a Distribution Market Environment" and "Power Flow Constrained Short-Term Scheduling of CHP Units".

The latest trends in the area of renewable energy integration are the self-consumption and islanding operation. Chapter "Microgrids Operation in Islanded Mode" investigates control and management issues in microgrids islanded operation mode, while Chapter "Islanding of Energy System" explores the islanding capabilities and benefits to increase the reliability of supply, especially in countries where infrastructure is still developing.

Overall, this book collects the latest broad, holistic and different aspects of energy systems integration for sustainable development. The importance of different resources and alternatives is discussed, to distinguish the advantages and challenges for each system. The integration of systems is also covered with detailed experiences and lessons from different regions and states in the effort to balance environment and development needs, and getting these to work in harmony with other social challenges. The complexity of multi-systems integration has given this book the approach to include some analysis to investigate the characteristics of smart integration. This book is intended for those working in the area of sustainable development in energy systems.

Paola, Malta Dr. Inġ. Eur. Ing. Brian Azzopardi

Acknowledgement

The editor wishes to thank the Malta Council for Science and Technology (MCST), Grants ENM-2016-002a Project ID eranetmed-11-286 3DMgrid Project—www.3dmicrogrid.com and ENM-2016-001 Project ID eranetmed_nexus-14-044 EDGE-WISE—www.edgewise.ubi.pt, through the ERANETMED initiative of Member States, Associated Countries and Mediterranean Partner Countries, and The MCST Internationalisation Partnership Award 2016 Grant IPAS-2016-015.

Contents

Overview on Microgrids: Technologies, Control and Communications 1
Antonio Carlos Zambroni de Souza, D.Q. Oliveira, Diogo Marujo, F.M. Portelinha, Jr. and B. De Nadai N

Hybrid Nanogrid Systems for Future Small Communities 19
Farhad Shahnia

Interconnected Microgrid Systems for Remote Areas 39
Farhad Shahnia

Distributed Agent-Based Coordinated Control for Microgrid Management 65
M.S. Rahman and A.M.T. Oo

Sustainable Development of Energy Systems in the Baltic Region 83
Jurijs Spiridonovs and Olga Bogdanova

Active Distribution Networks Operation Within a Distribution Market Environment 107
Geev Mokryani

Critical Performance Evaluation of a Few Intelligent Controllers for Effective Power Control in Variable Speed Wind Generators 119
Rajiv Singh and Asheesh Kumar Singh

Power Flow Constrained Short-Term Scheduling of CHP Units 147
Manijeh Alipour, Kazem Zare and Heresh Seyedi

Optimal Utilization of Solar Energy Resources in Hungary 167
Attila Talamon and Bálint Hartmann

Microgrids Operation in Islanded Mode 193
A.C. Zambroni de Souza, B. De Nadai N., F.M. Portelinha, Jr.,
Diogo Marujo and D.Q. Oliveira

Islanding of Energy System 217
Jayesh Joglekar

Overview on Microgrids: Technologies, Control and Communications

Antonio Carlos Zambroni de Souza, D.Q. Oliveira, Diogo Marujo, F.M. Portelinha, Jr. and B. De Nadai N

Abstract The modern society is highly dependent on electricity. At the same time, there is a huge concern about the greenhouse gas emissions and the current way of power generation, mainly with fossil fuels and high costs. The deployment of microgrids may be a solution for both questions. They are a solution to the supply reliability problem through the existent distributed energy resources (DERs) connected to the grid near customers, and the environmental concern is met by applying renewable sources to generate power. Beyond these goals, there is a supply cost reduction and revenue maximization through power trades with the main grid, which is also considered as desired matches. Using proper control strategies and robust communications techniques, microgrids generate, distribute and regulate the flow of power to consumers through a centralized or decentralized control. Smart microgrids are an ideal way to integrate renewable resources on the community level and allow for customer participation in all levels of the power market. Hence, a distribution system may evolve to a microgrid that may become a smart grid. Because the level of intelligence, communication, and control may vary, this chapter assumes the term microgrids in a general sense, which may stand for an ordinary microgrid or a smart grid. The goal of this chapter is to provide a brief overview on microgrids, including the state of art about the main motivation for the

A.C.Z. de Souza (✉) · F.M. Portelinha, Jr. · B. De Nadai N
Institute of Electrical Energy, Federal University of Itajuba, Itajuba, Brazil
e-mail: zambroni@unifei.edu.br

F.M. Portelinha, Jr.
e-mail: portelinha@gmail.com

B. De Nadai N
e-mail: nadaibruno@gmail.com

D.Q. Oliveira
Institute of Electrical Energy, Federal University of Maranhao, Sao Luis, Brazil
e-mail: dq.oliveira@ufma.br

D. Marujo
Department of Electrical Engineering, Federal University of Technology—Paraná, Medianeira, Brazil
e-mail: diogomarujo@utfpr.edu.br

B. Azzopardi (ed.), *Sustainable Development in Energy Systems*,
DOI 10.1007/978-3-319-54808-1_1

emergence of these networks. A survey on some existing microgrids around the world focusing on the different architecture and applications, e.g., hospitals, military bases and small isolated communities is carried out. Next, a typical microgrid's architecture is presented. Some other topics regarding generation resources, energy storage options, suitable communication technologies, and infrastructure are discussed. From the economic and social point of view, microgrid's effects are addressed by highlighting the main benefits. In addition, technical issues and future challenges are discussed in this chapter.

Keywords Microgrids · Renewable energy · Energy storage · Communication networks · Self-healing · Smart grids

1 Introduction

Electric power systems are traditionally divided into three main parts interconnected with each other: generation, transmission, and distribution. The *generation* concentrates the most generation sources, i.e., it represents the locations where the plants with larger capacities are connected. The *transmission* system is responsible for the interconnection between generation and loads, allowing the electrical system to operate optimally, both from a technical and economic point of view, under normal conditions or during emergencies. Finally, the *distribution* represents the final stage in the power transfer to the individual consumer. Previously, distribution systems were considered as passive and radial—the power flows in only one direction: from transmission to distribution system. In recent years, this scenario has been changing. Although it is not a new concept, the increasing use of distributed generation in the distribution system has transformed these networks [1].

According to [2], a large number of definitions about distributed generation are used. However, a general definition can be presented: "Distributed generation is an electric power source connected directly to the distribution network or on the customer side of the meter." Regarding the capacity of the generation power plants, no definitions have been made. Difficulties in defining the maximum capacity of distributed generation are found since every country has adopted values depending on the distribution system capacity and government regulations.

The integration of distributed generation influences electrical power systems conception. In particular, distribution networks are undergoing technological advances, remodeling the current network and allowing considerable changes in power system planning and operation. A key issue is related to the concept of active distribution networks (ADN). Active distribution networks are defined as networks with distributed generation in which the power flow can be bidirectional during some periods of the day [3].

If a part of the distribution system has enough generation capacity to supply its loads, energy storages devices, telecommunication and management infrastructure, this part of the system can operate isolated from the main grid, autonomously, forming a microgrid. The expression "Microgrids" was firstly presented in [4]. When the connection to the main grid is restored through the point of common coupling, the microgrid has the ability to export the surplus power to the main grid, i.e., it represents an active distribution network.

Several research projects on microgrids have been developed around the world. An important project on microgrids in the USA was created in 1999 but fully developed in 2002 [5]. Since it has been pursued by the Consortium for Electric Reliability Technology Solutions (CERTS), it has been called as "CERTS Microgrid." When it was created, this microgrid was composed by three 60 kW combined heat and power sources on a 160 m radial feeder of 480 V on the low-voltage side. Some other major projects in the USA and its characteristics are [6]:

- BC Hydro Boston Bar: 8.6 MVA hydroelectric generation connected to a feeder with 3 MW peak load;
- Mad River: 280 and 100 kW propane and biodiesel gensets, 30 kW microturbine and a photovoltaic array. The loads are composed of 12 residences and 6 commercial and industrial facilities.

An important project in Europe is called "More Microgrids: Advanced Architectures and Control Concepts for More Microgrids." Among other pilot microgrids, the following pilot plants have gained prominence [5–7]:

- The Kythnos Island Microgrid (Greece): 10 kW of photovoltaic array, 5 kW diesel genset and a nominal 53-kWh battery bank supply 12 houses;
- Continuon's MV/LV facility (Netherlands): 315 kW of photovoltaic array supplies 200 houses using four 400 m feeders;
- MVV Residential Demonstration (Germany): 30 kW of photovoltaic array provide energy to a residential area composed by 1200 inhabitant.

Japan stands out in Asia. According to [5], Japan is the current world leader in microgrid demonstration projects, since the government is encouraging the use of renewable sources, such as solar and wind power. The main Japanese microgrids are Aomoni, Aichi and Kyoto. The latter is composed by 50 kW of photovoltaic array, 50 kW of wind power, 5×80 kW biogas gensets and 250 kW of ER-molten carbonate fuel cell and 100 kW battery bank, covering an area of 40 km.

Beyond this initial approach to the historical evolution of the emergence of microgrids, an overview of the subject throughout this chapter is presented. Firstly, typical microgrid architecture is shown, in such a way that the main devices are briefly discussed. Next, power management and communications requirements are pointed out. The key features and requirements of microgrids are then shown.

2 Microgrid Architecture

As presented in the previous section, several microgrids topologies can be found around the world. These networks may present different structures due to several factors, but mainly according to the network type (AC or DC), the topology and the generation sources available. However, by definition, a microgrid must be formed by generation sources, energy storage systems, and loads. Based on [8], a generalized AC microgrid comprising a large number of devices is shown in Fig. 1. In the figure, the dashed and continuous lines are the information and electric power flow, respectively.

According to Fig. 1, an AC microgrid is composed by three feeders connected to the main grid. The low-voltage side is connected to the high side through a transformer. The point where a microgrid is connected to the main grid is the point of common coupling (PCC). A circuit breaker (CB) allows decoupling the microgrid from the main grid in cases of disturbances, changing the operation from grid-connected mode to islanded mode. There may be other intermediate circuit breakers that divide the microgrid in nanogrids. Although the main part of the microgrid shown in Fig. 1 is in low voltage, there are medium-voltage microgrids, as in the case of rural microgrids.

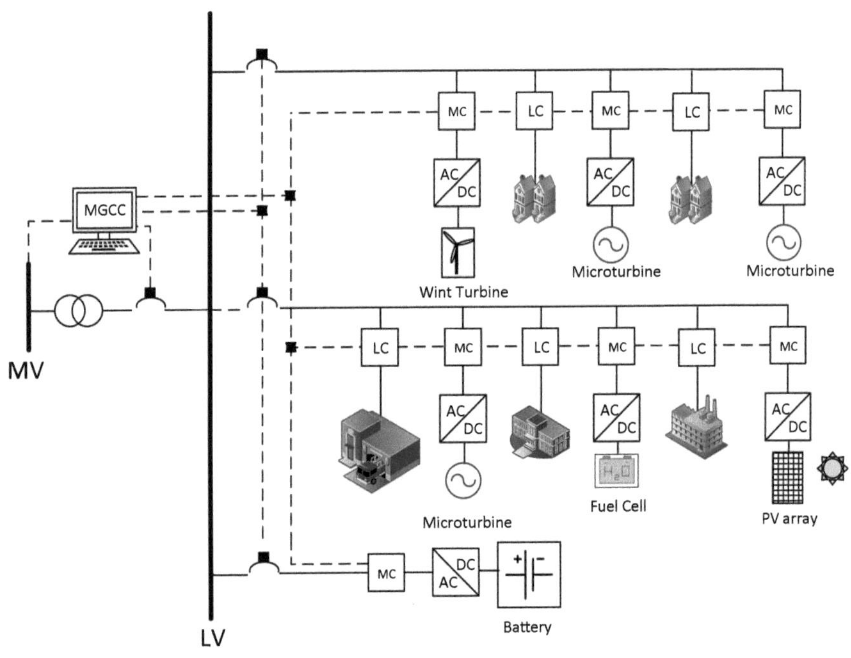

Fig. 1 Generalized AC microgrid structure

In Fig. 1, dispatchable and non-dispatchable generation sources are available. In dispatchable sources, the generated power amount can be controlled, i.e., the primary fuel can be stored. Some examples of this kind of source are microturbines and fuel cells. In contrast, the non-dispatchable source depends on the intermittent renewable sources such as solar (photovoltaic panel) or wind (wind turbine). These sources are interfaced with the help of power electronic converters (PEC). Although these converters allow operating flexibly, no additional inertia is added to the system from the distributed generators [9]. This low inertia can affect the system frequency, mainly when disconnected from the main grid. In the islanded mode, the energy storage units compensate the imbalance between generation and load; in the connected mode, the swing bus has the responsibility for balancing load/generation to ensure constant frequency and loss supply.

The operation and management of microgrid are controlled through local and central controllers [3]. As depicted in Fig. 1, local controllers are represented by load controller (LC) and microsource controller (MC). The microgrid central controller (MGCC) coordinates the microgrid operation globally. Some of the main functions of these controllers are highlighted in [3, 9] and listed as follows:

- Load controller: Perform control in controlled loads by connection/disconnection of certain equipment in predetermined periods. LC also relieves unfavorable operating condition of microgrid;
- Microsource controller: This controller uses local information to control the microsource during disturbances. Besides the connection/disconnection of generation sources in the microgrid, MC main function is to regulate the power flow, voltage, and frequency on a feeder where the operating point varies continuously;
- Microgrid central controller: It can be considered as the microgrid's "brain." The MGCC is responsible for determining and sending the set points to the LC and MC to maintain voltage and frequency within specific ranges. The latter controllers connect/disconnect the load or change the voltage/active power according to the set point determined by the MGCC. Through measurements, the MGCC has the responsibility to coordinate other functions: load forecasting, scheduling of generation, security assessment and demand side management. Another function related to MGCC is to control the energy storage devices. The batteries must be charged at times with excess generation. When generation is missing, i.e., the load is bigger than the available generation, batteries should assist the operation by supplying power to the microgrid. More details about the MGCC functions are linked with the power management framework, which is addressed in the next section.

3 Distributed Energy Resources

Distributed energy resources (DERs) are considered modular and small-scale resources which could be plugged into the utility grid or directly into the costumer's installation to provide local supply of electricity [10]. DERs are strongly present in the microgrids concept.

In this section, the DERs are classified in three categories: dispatchable, non-dispatchable generation and energy storage systems (ESS). A brief discussion about each one is presented as follows.

3.1 Dispatchable Generation

As stated before, dispatchable units have the power output controlled by the MGCC. This kind of generation is responsible for power balancing and voltage control since the primary fuel is available at any time and it is possible to control the power injection and voltage output.

Some examples of dispatchable generation are microturbines, fuel cells, gensets and small hydroelectric plants.

3.2 Non-Dispatchable Generation

Non-dispatchable generation includes the renewable-based plants. Differently from dispatchable units, this kind of generation cannot have controlled power output as a consequence of the volatility and intermittence of weather conditions that modify the availability of energy throughout the day [10]. In this case, the injected power is related to the availability of the primary source, regardless of the voltage and frequency control. The solar and wind generation are examples of non-dispatchable sources.

3.3 Energy Storage Systems

The ESS can operate as load or generator, depending on the balance between generation and load. If the frequency is lower than the nominal, the ESS injects power into the system until the capacity is exhausted; if the frequency is bigger than the nominal, the ESS absorbs power from the grid. These devices act as controllable sources with very fast output, e.g., low time constant of response, to face sudden system changes, mainly in load-following situations. Thus, they are the first corrective mean to restore frequency [8].

The batteries are the main option in ESS for application in microgrids; however, others examples have also been applied, such as flywheels, compressed air and pumped hydroelectric.

It is not usual the direct connection of DERs in microgrids. This is related to the characteristics of the generated power, such as the low inertia, the frequency randomness of the primary source and, in some cases, the application of direct current technology in generation. In this sense, the DERs are connected through an electronic interface, i.e., via PEC. These converters act with defined strategies to control the pulse-width modulation (PWM) in the semiconductors bridges and, consequently, the power, the voltage, and the frequency outputs. The operation mode of each one of these DERs is defined by power management strategy adopted by MGCC [8]. The control strategies of PECs will be discussed in the next section.

4 Power Management

Microgrids operation is quite different from the standard power distribution systems due to the presence of small-scale renewable power generation, critical and controllable loads, distributed energy storage and so on.

The power management approach should be aware of these differences, which include but are not limited to [11]:

- steady-state and dynamic features from DERs, especially those coupled via electronic interface;
- the intermittence from the primary sources in each site;
- planning and management of energy storage units;
- microgrid's current status, i.e., grid-connected or islanded operation;
- the quality of power and the presence of high priority loads that demand preferential service.

As the microgrid is designed to be an autonomous system, the operation must be supported by a power and energy management system (P/EMS) and some smart features are expected to be present. The P/EMS is responsible for [11]:

- managing the different DERs connected to the grid;
- assessing/monitoring microgrid's frequency and nodal voltages as far as other power quality indices;
- planning/operating the microgrid in standalone and emergency conditions;
- deciding the moment to connect and disconnect the microgrid from the main grid;
- considering operation optimization by internal and external data;
- improving dynamic response, maintaining stability and nominal values for state variables.

The P/EMS is responsible for assessing many variables by gathering data that come from the advanced metering infrastructure (AMI) through a reliable communication network. Considering that all data from generation and demand are available, the P/EMS can: update the reference values for active power P_i, where i is the electronically coupled unit; take demand management actions by connecting and disconnecting lower priority customers; optimize the power generation aiming to save resources and to minimize the power generation costs and power purchases from the main grid or to maximize power exportation to increase profits in energy market; decide whether is the best moment to connect/disconnect the microgrid according to current energy prices and available generation resources.

The P/EMS may be accomplished using a centralized or a distributed approach. In the centralized approach, it runs in a MGCC which collects all data and sends the assessed variables through the bidirectional communication network. This communication network should be reliable and robust, and depending on the amount of data flow, it should have a big capacity.

The centralized approach also demands a powerful processor, robust and fast algorithms to perform calculations in a timely manner, since these data must be sent back. The results assessed in these conditions are optimal since all necessary data inputs are available for calculations. Because the microgrid is located in a small geographical area, the centralized approach is still possible due to the smaller amount of data and time delays for communication signals.

The distributed approach demands the existence of many local controllers in the microgrid. These controllers, called previously as LCs and MCs, can be called agents. They are responsible for assessing local variables and communicate with their neighbors to transfer data and initiate requests. These agents are organized in different layers, where they can communicate to others agents in the same layer and to their superior agent in the layer above. The superior agents are responsible for collecting data from their subordinate agents and assess more general variables. The results are informed to each agent in the inferior layer. This framework is kept in all microgrids until the most superior layer, which also communicates with the external world.

This decentralized approach is possible since the PEC from renewable generation units and many different physical components have powerful microcontrollers that assess local variables.

The power management in microgrids can also be investigated considering the two different operation modes: connected to the grid or standalone/islanded mode. This is possible since these are completely different scenarios which demand different approaches due to the respective requirements.

In connected mode, the microgrid operates connected to the main power grid. This grid is more robust and may be considered as a slack bus. This means that the main grid produces the frequency reference necessary to the microgrid's DERs synchronization and it also assures the power balance. The microgrid's dynamic behavior is also affected since the main grid has a higher inertia, enabling a lower frequency variation in case of disturbances.

During the operation in connected mode, the P/EMS optimization algorithms may pursue some of these objectives, but are not limited to:

- minimization of power losses;
- minimization of power generation costs and carbon emissions;
- maximization of revenue by exporting power to the main grid;
- maximization of power quality indices.

Ancillary services to the main grid are also possible, including reactive power support and black start.

In the islanded mode, the microgrid operates disconnected from the main grid. This is a challenging situation since the biggest issue now is to provide a frequency reference to the DERs aiming the microgrid synchronization and voltage control. To accomplish this goal, some PEC from electronically coupled units should operate as voltage source inverters (VSIs) to provide voltage and frequency references to the other DERs by emulating the behavior of a synchronous machine. This approach is called Multi-Master Operation.

Some references in the literature [8, 12] suggest using a big energy storage unit as a slack bus for the microgrid due to the fast response and the capacity to meet other requirements, such the power balance due to high variations in renewable generation. In practice, this idea may be accomplished by the different droop configurations from the DERs. The other inverters should operate in PQ mode, supplying defined set points for active and reactive power.

In the islanded operation scenario, the P/EMS has different goals from the connected operation mode. This is due to the different requirements from the islanded operation, which includes but are not limited to:

- Minimization of high priority load shedding;
- Minimization of switching operations, in case of network reconfiguration due to faults or to keep the supplying of higher priority customers;
- Maximization of supplying time for the highest priority customers, considering the availability of power.

The islanded operation should include a load shedding scheme, which is continuously updated by the P/EMS according to the available data from the power generation units and customers.

5 Communications

Communications play a crucial role in the context of smart grids [13]. Many applications have been researched, but only a few real ones have been developed [14]. The main required features in a smart grid/microgrid are self-healing, demand management, protection against physical and cyber-attacks, distributed energy resources, increased energy quality and more customer participation in the energy market [15]. Some of these requirements are possible if a reliable and robust communications network is implemented, where information flows in a bidirectional way, providing more customers and machine interaction with the grid. The information in real time will allow isolating affected areas, to redirect the power flow and to maintain the largest possible number of consumers supplied by helping to prevent interruptions [16]. The machine-to-machine type of communication (M2M) [17] describes better the microgrid scenario, where machines communicate with each other without any human interaction. The massive introduction of sensors and automated controls make possible to anticipate, detect and take decisions autonomously [18].

For this to happen, a communications system must be implemented and offer low risk of attack, resiliency, robustness and a low-cost infrastructure [19]. In the next sections, a brief understanding of the communications requirements and the possible technologies to be implemented is presented. The purpose of this section is to provide to the reader a brief overview on the communications challenges and issues in microgrid operation.

5.1 Communications Overview for Microgrids

The concept of integrated communications in smart grids provides better communication between all system components. The networks will require real-time communication in many cases. The Information and Communications Technology (ICT) infrastructure must be reliable, highly available, scalable, secure and easy to manage.

The concept of smart grid communication involves multiple entities with potential data exchange that can reach large volume of data and different response times [20]. The performance should be scalable to allow entry of new devices. Thus, it is necessary to understand the requirements for communication between devices and to outline how it would be possible to build a microgrid communication network. Each microgrid will have a limited number of smart meters ranging from a few hundred to a few thousand linked to some substation control centers. Some very important requirements for the operation of communications networks are [21]:

- Latency: It is the delay in the network or how long it takes for a packet to travel from one point to another in the network;
- Bandwidth: It means the maximum amount of data which can be sent in a network, measured in bit per second;
- Data rate: maximum data transmission rate, which varies upon the chosen technology;

Table 1 Proposed microgrids layers

Application	SCADA	Distribution grid management	Energy resources and storage	Demand response	Electric transportation	Advanced metering infrastructure
Availability (%)	99–99.99	99–99.99	99–99.99	99–99.99	99–99.99	99–99.99
Latency	20–200 ms	100 ms–2 s	20 ms–15 s	500 ms-several minutes	2 s–5 min	2–15 s
Data rate	9 kbps	9–100 kbps	9–56 kbps	14–100 kbps	0.1–2 Mbps	500 kbps
Reliability	High	High	High	High	High	Medium
Security	High	High	High	High	High	High

- QoS—quality of service: It measures the overall performance of the system and guarantees the quality of a service, measuring elements such as bandwidth, latency and error rate.

The smart grid integrates advanced ICT, automation, sensing, and measurement technologies to achieve these objectives. This network must be integrated with high performance, high reliability, scalability and ubiquitous. The communication network will be responsible for collecting and forwarding data, monitoring of all devices and act according to the received data. Table 1 illustrates a brief review of all requirements and their aspects in microgrid communications. The data were retrieved of all major publication on smart grid communications requirements [22].

5.2 *Communications Architectures*

Communication networks are composed of layers, depending mainly on its size and data communications requirements. Microgrid architecture can be described in three layers according to the design operation. Figure 2 illustrates the design model [23].

The microgrid layers can be described as home area network (HAN), neighborhood area network (NAN) and wide area network (WAN).

The HAN covers a very small geographical area. Its applications are mainly customer automation. The communications requirements in this level are low. The possible communication technologies that can be applied are Bluetooth, ZigBee, WiFi, Ethernet and power line communication (PLC).

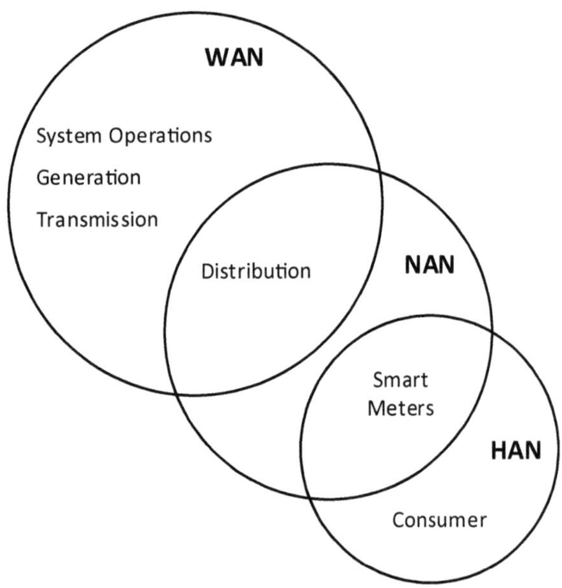

Fig. 2 Microgrid network

Table 2 Layer requirements

Microgrid layer	Coverage range
Wide area network (WAN)	Long distances
Neighborhood area network (NAN)	Medium distances
Home area network (HAN)	Small distances

The role of NANs in a microgrid context is more complex. The applications in this layer are demand response, distribution automation, smart meters. As a large number of devices from consumers will be connected, the data requirements will increase and more robust technologies must be applied.

Wide area awareness, real-time protection, and self-healing are the main tasks of WAN, which requires a very robust and reliable technology. Latency times can be very low, so their time to respond to any failure must be fast. Several communications technologies can be applied to cover this layer requirement, such as wire and wire less technologies.

Table 2 illustrates the coverage and minimum communications data rates for each layer.

The use of known and new telecommunications technologies to meet various applications scenarios provided by the operation of microgrid must consider regulatory aspects, technical feasibility, reliability and investment to build an infrastructure. In the next section, a brief overview of all detailed requirements to cover all microgrid layer applications will be properly discussed.

5.3 *Wire and Wireless Technologies Candidates*

Many technologies are available for data transmission between the consumer unit and the operation center, and from the utilities to the whole system. The choice of a communication technology should be based on the need for reliability, security, and availability of each service offered. The aim is to provide an overview on the current state communication technologies, giving some key directions to choose a communication system and how it can be interconnected with the distribution system, considering its characteristics, applicability and availability.

The type of operation to be performed is another issue that should be taken into account. Those considered critical, such as control and key operation will require a more robust infrastructure, very low latency response and must be available all time.

The objective is to deliver an intelligent electrical system which monitors, protects and control the grid without the need of human interaction. The communication infrastructure must have a very low investment, otherwise can be considered unviable by the stakeholders. Thus, existing technologies must be considered as solutions, decreasing the implementation costs and enabling the

Table 3 Overview of telecommunication technologies for microgrids

Proposed technology	Data rate	Coverage range	Advantages	Disadvantages
PLC	NB-PLC 1-500 kbps	150 km or more	Low cost of deployment	Interference causes signal attenuation
	Home plug—200 Mbps	Up to 3 km		
Coaxial cable	up to 170 Mbps	Up to 25 km	Low cost of deployment	Suffer influence of noise
Optical fibers	PON (passive optical network) up to 1 Gbps	EPON (Ethernet passive optical network) up to 20 km	Robustness against noise, achieve long distances	High deployment cost
	SONET—Synchronous Optical Networking / SDH—Synchronous Digital Hierarchy 40 Gbps	Up to 100 km		
	WDM 40 Gbps	Up to 100 km		
WiFi	Up to 600 Mbps	Up to 1 km	Low cost of deployment, easy to implement	Suffer high interference, need of licensed spectrum
ZigBee	Up to 250 Kbps	Up to 100 m	Low cost of deployment, easy to implement	Low data rate
Bluetooth	Up to 720 Kbps	Up to 100 m	Low cost of deployment, easy to implement	Low data rate
WiMax	Up to 1 Gbps	Up to 10 km	Achieve long distances, support of millions of users	High deployment cost, need of licensed spectrum
2G/3G/4G	Up to 86 Mbps—dependency on the chosen technology	Up to 100 km	Low infrastructure cost, high data rate, support of millions of users	Use of licensed spectrum, cost per user
Satellite	Up to 450 kbps	Large geographical areas	Achieve long distances, high reliability link	High deployment cost, not suitable for low latencies scenarios

application in microgrids. Some technologies are presented in Table 3. Note that not all technologies available are listed.

Table 3 gives an overview on the communication systems available in the market listing their current features. A previous study on microgrid requirements pointed some technologies as possible candidates [24–35]. Each possible technology is described according to latency, data rate, QoS and coverage features. Advantages and disadvantages are also summarized in Table 3. The possible solution for an optimal communication network can be a wired or a wireless technology, or a hybrid solution.

6 Benefits, Issues and Challenges

The microgrids will be the most promising evolution in electrical power systems. The strong and intelligent integration among all agents of these systems makes an environment very different from the electrical power system current scenario, highlighting, mainly, the massive presence of control and telecommunication network and data processing. These features increase the reliability, security and quality of electric service for any appliance to microgrids.

From technical and social points of view, some important benefits of this new concept are listed as follows [36]:

- The improvement in grid reliability by inserting controlled distributed generation at distributed level, near the consumer;
- The contribution in the reduction in greenhouse gases emissions by considering the integration of renewable resources;
- Ability to reach energy efficiency goals;
- The participation of the consumers in the wholesale energy market;
- The possibility to manage the consumer's energy cost by changing the penetration of renewable and non-renewable resources.

Furthermore, resiliency, reliability, emission reduction, reduced costs of recurring system upgrades, enhanced energy efficiency and power quality and lower energy costs are benefits also present in the concept of microgrids [10].

Even though microgrids have a large number of benefits, the path to popularize this type of electric power system will not be easy. There are still many challenges that must be overcome, mainly, concerning control and operation aspects.

For remote areas, the microgrids have already become a possible solution to supply small communities, where there is no distribution system while the energy supply is a result of local and renewable resources, such as solar, wind and biomass or biogas. On the other hand, the creation of microgrids in urban communities connected to utility grid has found some technical and environmental obstacles.

A complex change in the traditional infrastructure is necessary in order to transform active distribution systems in microgrids. When generators are plugged into a distribution system, the power flow that previously was unidirectional changes to bidirectional. This requires a new adjustment on the topology of electrical utility in planning and operation aspects, demanding investments which causes, in many times, resistance by electrical utility. Another topic that should be considered is the fact the operation of power system has not been planned to operate with large presence of distributed generation.

In summary, some points of risks regarding microgrids implementation must be considered to its development [36]:

- The consumer changes to the DER concept, becoming a new active agent in the energy market;
- The absence of a large-scale control infrastructure;
- Scheduling the introduction of DERs at each customer site, considering de demand response;
- Policies of incentive and necessary capital markets for investors;
- The utility grid needs to create new policies to manage distributed generation, besides to educate and coordinate the regulatory agencies.

The presence of MGCC allows microgrids the ability to provide load and source control in a way to guarantee a safe and optimized operation; telecommunication network which interconnects all agents is the most important challenge and which characterizes a microgrid. The presence of MGCC replaces the operator, enabling that the decisions are made in autonomous way based on priority loads, weather conditions, power losses and energy prices in economical market or any kind of entry, so the MGCC is considered the brain of the microgrid. If the system cannot operate within safe limits in islanded mode, it is not a microgrid.

Nowadays, the efforts of researches about this issue are related to attempt to standardize the MGCC and its all components. Defining protocols of communication, policies of load shedding, control hierarchy, and synchronism strategies among others in the microgrids level constitute an important step in direction to modernize electrical power systems. In the future, the traditional and defined hierarchical generation, transmission, distribution and consumer of electrical energy will be substitute by an environment totally integrated in all levels.

References

1. J. Machowski, J.W. Bialek, J.R. Bumby, "Power System Dynamics and Stability" John Wiley & Sons, 1997.
2. T. Ackermanna, G. Andersson, L. Söder, "Distributed generation: a definition," *Electric Power Systems Research*, vol. 57, pp 195–204, 2001.
3. S. Chowdhury, S. P. Chowdhury, P. Crossley, "Microgrids and Active Distribution Networks," IET Renewable Energy Series 6, 2009.

4. R.Lasseter "Microgrids," IEEE Power Engineering Society Winter Meeting Conference Proceedings, pp. 305–308, 2002.
5. N. Hatziargyriou, H.Asano, R. Iravani, C. Marnay "Microgrids: An Overview of On going Research, Development, and Demonstration Projects" *IEEE power & energy magazine*, pp. 78–94, 2007.
6. M. Barnes, J. Kondoh, H. Asano, J Oyarzabal, G. Ventakaramanan, R. Lasseter, N. Hatziargyriou, T. Green, "Real-World MicroGrids-An Overview," IEEE International Conference in System of Systems Engineering,, vol., no., pp. 1–8, 16-18 April 2007.
7. More Microgrids: Advanced Architectures and Control Concepts for More Microgrids, available on: http://www.microgrids.eu/index.php?page=kythnos
8. J. A. P. Lopes, C.L. Moreira, A.G. Madureira, " Defining control strategies for microgrids islanded operation," *IEEE Transactions on. Power Systems*, vol. 21, no. 2, pp. 916–924.
9. M. S. Mahmoud, M. Saif Ur Rahman, F. M. A. L-Sunni. "Review of microgrid architectures —a system of systems perspective," *IET Renewable Power Generation,* vol. 9, no. 8, pp. 1064–1078, 2015.
10. S. Parhizi, H. Lotfi, A. Khodaei, and S. Bahramirad, "State of the Art in Research on Microgrids : A Review," *IEEE Access*, vol. 3, Denver, pp. 890–925, Jul-2015.
11. F. Katiraei, R. Iravani, N. Hatziargyriou, A. Dimeas, "Microgrids Management," *IEEE Power & Energy Magazine*, pp. 54–65, May/June 2008.
12. C. L. Moreira, F. O. Resende, J. A. Peças Lopes, "Using Low Voltage MicroGrids for Service Restoration," *IEEE Transactions on Power Systems,* vol. 22, no. 1, pp. 395–403, 2007.
13. K. Tomsovicand D. E. Bakken, "Designing the Next Generation of Real-Time Control, Communication, and Computations for Large Power Systems". IEEE Proceedings, V. 93, No. 5, pp. 965–979, 2005.
14. Z. Fan, P. Kulkarni, S. Gormus, C. Efthymiou, G. Kalogridis, M. Sooriyabandara, Z. Zhu, S. Lambotharan, and W. H. Chin, "Smart Grid Communications : Overview of Research Activities" vol. 15, no. 1, pp. 21–38, 2013.
15. J. Frolik, a L. Lentine, a Seier, and C. Palombini, "for μGrid Agents," 2012 3rd IEEE PES Innov. Smart Grid Technologies Europe, pp. 1–5, 2012.
16. V. C. Güngör, D. Sahin, T. Kocak, S. Ergüt, C. Buccella, C. Cecatiand G. P. Hancke, "Smart Grid Technologies: Communication Technologies and Standards". IEEE Transactions on Industrial Informatics, vol. 7, no. 4, pp. 529–539, 2011.
17. S. K. Tan, M. Sooriyabandara, and Z. Fan, "M2M Communications in the Smart Grid: Applications, Standards, Enabling Technologies, and Research Challenges," *International Journal of Digital Multimedia Broadcasting*, vol. 2011, pp. 1–8, 2011.
18. E. Ancillotti, R. Bruno, and M. Conti, "The role of communication systems in smart grids: Architectures, technical solutions and research challenges," *Computer Communications*, vol. 36, no. 17–18, pp. 1665–1697, 2013.
19. R. Bi, M. Ding, and T. T. Xu, "Design of Common Communication Platform of Microgrid," pp. 735–738, 2010.
20. V. C. Gungor, D. Sahin, T. Kocak, S. Ergut, C. Buccella and C. Cecati, G. P. Hancke, "A Survey on Smart Grid Potential Applications and Communication Requirements". IEEE Transactions on Industrial Informatics, vol. 9, no. 1, pp. 28–42, 2013.
21. P. Kansal, A. Bose, "Bandwidth and Latency Requirements for Smart Transmission Grid Applications". IEEE Transactions on Smart Grid, vol. 3, no. 3, pp. 1344–1352, 2012.
22. X. Fang, S. Misra, G. Xue and D. Yang, "Smart Grid—The New and Improved Power Grid: A Survey". In *IEEE Communications Surveys & Tutorials*, vol. 14, no. 4, pp. 944–980, Fourth Quarter 2012.
23. M. Kuzlu, M. Pipattanasomporn, S. Rahman, "Communication network requirements for major smart grid applications in HAN, NAN and WAN," *Computer Networks*, vol. 67, pp. 74–88, 2014.
24. Y. Yan, Y. Qian, H. Sharif, and D. Tipper, "A Survey on Smart Grid Communication Infrastructures : Motivations, Requirements and Challenges". IEEE Communications Surveys & Tutorials, pp. 1–16, 2012.

25. M. Daoud and X. Fernando, "On the Communication Requirements for the Smart Grid". *Energy and Power Engineering* vol. 3, no. 1, pp. 53–60, 2011.
26. S. Galli, A. Scaglione and Z. Wang, "For the Grid and Through the Grid: The Role of Power Line Communications in the Smart Grid". Proceedings of the IEEE, vol. 99, no. 6, pp. 998–1027, June 2011.
27. S. Galli, A. Scaglione and Z. Wang, "Power Line Communications and the Smart Grid". IEEE International Conference on Smart Grid Communications (SmartGridComm), Gaithersburg, pp. 303–308, 2010.
28. P. Lv, X. Wang, Y. Yang and M. Xu, "Network Virtualization for Smart Grid Communications". *IEEE Systems Journal*, vol. 8, no. 2, pp. 471–482, June 2014.
29. B. Akyol, H. Kirkham, S. Clements, and M. Hadley, "A survey of wireless communications for the electric power system". US Department of Energy, January, 2010.
30. B. Ran, E. Negeri, N. Baken, and F. Campfens, "Last-Mile Communication Time Requirements of the Smart Grid". In: Sustainable Internet and ICT for Sustainability, pp. 1–6, 2013.
31. S. Clements, T. Carroll, and M. Hadley, "Home Area Networks and the Smart Grid". US Department of Energy, 2011.
32. M. R. Souryal and N. Golmie, "Analysis of advanced metering over a Wide Area Cellular Network". *IEEE International Conference on Smart Grid Communications (SmartGridComm),* Brussels, pp. 102–107, 2011.
33. R. Munoz *et al.*, "The CTTC 5G End-to-End Experimental Platform: Integrating Heterogeneous Wireless/Optical Networks, Distributed Cloud, and IoT Devices". *IEEE Vehicular Technology Magazine*, vol. 11, no. 1, pp. 50–63, March 2016.
34. M. Simsek; A. Aijaz; M. Dohler; J. Sachs; G. Fettweis, "5G-Enabled Tactile Internet," *IEEE Journal on Selected Areas in Communications*, vol. 34, no. 99, pp. 460–473,2016.
35. D. Chang, J. Lee and T. H. Lin, "Smart satellites in Smart Grids," *IEEE International Conference on Smart Grid Communications (SmartGridComm),* Venice, 487–492, 2014.
36. T. Mohn, "Campus microgrids: Opportunities and challenges". In: IEEE Power and Energy Society General Meeting, San Diego, CA, pp. 1–4, 2012.

Hybrid Nanogrid Systems for Future Small Communities

Farhad Shahnia

Abstract The power supply system of future communities can be considered in the form of a small microgrid or a nanogrid which is mainly based on renewable energy resources. Such a system can lead to a sustainable development of electrical systems and can help the current and future generations to access the benefits of electricity without adding more emissions and pollutions to the environment. A nanogrid should have adequate generation capacity in its distributed energy resources (DERs) to supply its demand in the off-grid status. It should also be able to exchange power with an existing utility feeder. The operation principle of a nanogrid is discussed in this chapter. A hybrid nanogrid may consist of one ac bus and one dc bus, which are connected via a power electronics-based converter. Some DERs and loads of the community will be connected to the ac bus of the nanogrid, while some DERs and loads will be connected to its dc bus. The converter facilitates power exchanges among the buses and also controls their voltages, during off-grid operation. The dynamic performance of a small community nanogrid is discussed in detail here.

Keywords Distributed energy resource · Nanogrid · Future communities

1 Introduction

Sustainable development is the core principle of maintaining our finite resources that are necessary for our future generations. Regarding electrical systems, sustainable development strongly depends on the application and control of renewable energy resources for electricity generation instead of the fossil fuels such as coal or gas. Also, it is highly desired to generate the electricity near the consumers such that minimum adverse impact is imposed on the nature when building long transmission and distribution lines. Environmental concerns, high electricity bills, and

F. Shahnia (✉)
School of Engineering and Information Technology, Murdoch University, Perth, Australia
e-mail: f.shahnia@murdoch.edu.au

B. Azzopardi (ed.), *Sustainable Development in Energy Systems*,
DOI 10.1007/978-3-319-54808-1_2

the financial incentives that are in place by the governments motivate the householders to install renewable energy-based distributed energy resources (DERs) and energy storage systems in their premises to locally generate and store energy [1, 2]. With this trend, in near future, the majority of householders are expected to have some types of DERs that are connected to the utility feeder. Thus, merging all the DERs of a group of neighboring householders, it is possible to envision the formation of a small-scale microgrid, referred to as nanogrid [3, 4]. Thus, these neighboring households, denoted in this chapter by community houses, can collectively consume the generated power of the solar photovoltaic-type DERs and battery-type energy storages in grid-connected as well as off-grid statuses.

The majority of the present residential electric appliances work with ac voltage; thus, power electronics-based systems are needed for converting the dc voltage that is generated by photovoltaic-, wind- and battery-type DERs to the ac voltage. Recently, a dc voltage is suggested to supply certain types of residential loads such as lights, TVs, computers directly [5]. A comparison shows that the efficiency of supplying electricity to the residential appliances with a dc voltage of dc generators is greater than the efficiency of supplying them with a converted ac voltage [6, 7]. This improved efficiency is due to the reduction of the power electronics-based conversions. As an example, dc loads such as charging electric vehicles can be supplied directly from the dc bus, eliminating the ac–dc converters. Similarly, the first stage of ac–dc conversion in variable speed drives can be removed with an existing dc bus. This will result in reduced power conversion losses as well as the required power for cooling and ventilation. Additionally, a dc system has reduced copper losses due to lower effective resistance in the conductors, owing to the absence of skin effect and the reactive power flow in the cables [8]. These facts contribute to the increased efficiency of a hybrid ac–dc nanogrid.

Since 2002 that the microgrid concept was introduced [9], numerous researches have been carried out in the area of ac microgrids [10–15]. In recent years, the dc microgrids have also gained the attention of the researchers, thanks to their better efficiency and flexibility in their control. The hierarchical control system and an adaptive droop control of a dc microgrid system are introduced in [16–19], while the control of the DERs in a dc microgrid is studied in [20, 21]. The power electronic interface to connect a dc bus with an ac bus of an MG is introduced in [22], while a current-controlled approach has been utilized in [23] to control the power flow and the power factor of such bidirectional converters.

2 Hybrid AC–DC Nanogrid Structure

With the above review on the existing techniques, this chapter focuses on a limited-size, low-voltage network, denoted by nanogrid, which is assumed as the electric supply system of future small communities. Figure 1 illustrates schematically such a network. The considered nanogrid integrates dc and ac DERs and loads, which are linked to a dc and an ac bus. These two buses are then interconnected through a power electronics-based converter, referred to as tie-converter. The nanogrid can also exchange power with the utility feeder according to the generation of the DERs. However, it is also capable of operating in off-grid status. The ac bus is also interlinked to the utility feeder via a static switch (SS) and a three-phase power transformer. The ac and dc buses are required to operate individually and independently. However, if there is an increase in power demand or a shortage of power generation which causes voltage deviation in the dc bus or voltage/frequency deviation in the ac bus, the tie-converter takes action to regulate the voltage and frequency.

The AC bus is thought to be a 415-V three-phase system and is assumed to be fed by a 30-kVA transformer, in this chapter. Two DERs are thought to be connected to the ac bus through power electronic converters, while another two power electronic-interfaced DERs are thought to be linked to the dc bus. In this study, it is also supposed that a boosting transformer with a turns ratio of $a = 2$ is used in the structure of the tie-converter. Since uV_{dc} needs to be about 1.35–1.6 of the line voltage of the three-phase system, V_{dc} has been selected as 350 V [24]. If a higher turns ratio is chosen for the transformer, a lower dc voltage can be selected.

The nanogrid should be controlled such that it operates dynamically in the following modes [25]:

Mode-1: isolated and independent ac and dc buses while the ac bus is isolated from the utility feeder,
Mode-2: isolated and independent ac and dc buses while the ac bus is coupled to the utility feeder,
Mode-3: interconnected ac and dc buses while the ac bus is isolated from the utility feeder,
Mode-4: interconnected ac and dc buses while the ac bus is coupled to the utility feeder,

These modes are shown schematically in Fig. 2.

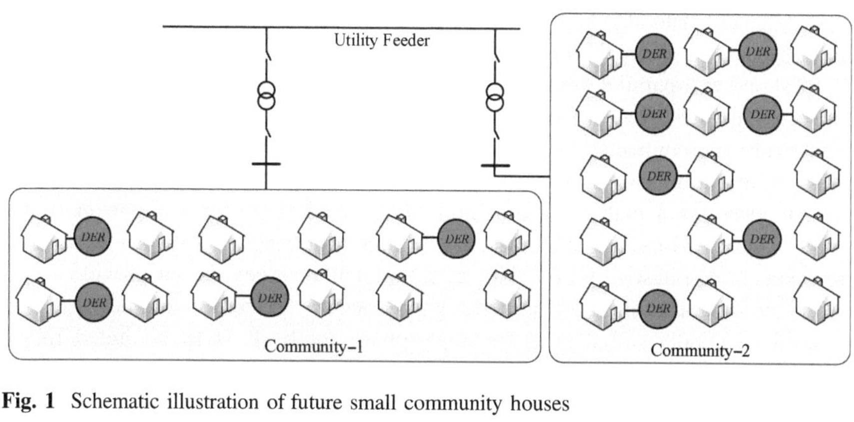

Fig. 1 Schematic illustration of future small community houses

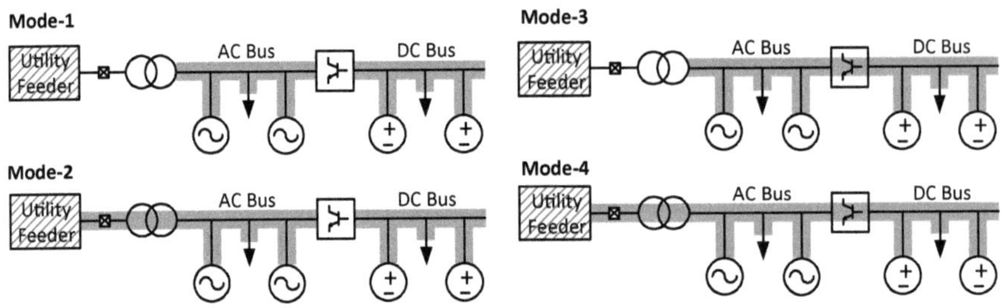

Fig. 2 Different modes of operation of the considered hybrid ac–dc nanogrid

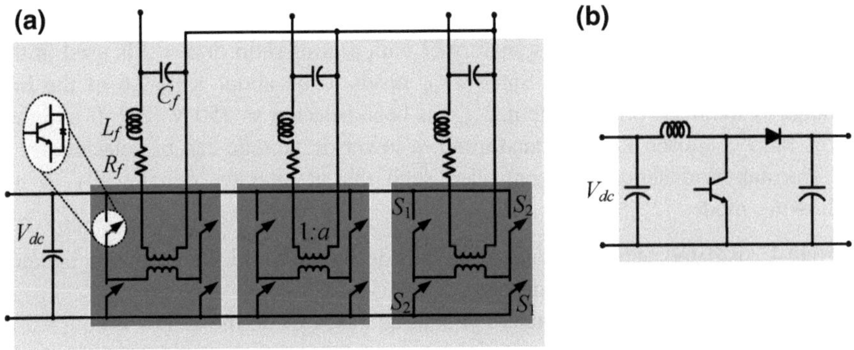

Fig. 3 **a** Considered VSC and filter structure for DERs of ac bus, **b** considered boost converter structure for DERs of dc bus

2.1 AC Bus

A group of DERs and loads form a three-phase ac section. Every DER is coupled to this bus via a three-phase, voltage source converter (VSC) and an LC output filter, as shown schematically in Fig. 3a. Every VSC comprises of three, single-phase H-bridges, while every H-bridge consists of the insulated gate bipolar transistors (IGBTs) and anti-parallel diodes. A single-phase transformer, with 1:a turns ratio, is connected to the output of every H-bridge. The secondary sides of these transformers are star-connected. The transformers offer galvanic isolation for the DERs. This configuration can be utilized to supply single-phase and unbalanced loads since it provides a path for circulation of the zero sequence component of the current. Alternatively, a three-phase, 4-wire (neutral-clamped) VSC can be utilized. However, three-phase, 3-wire VSCs are not suitable as they do not provide a path for the circulation of the zero sequence component of the current and hence cannot be used to supply unbalanced or single-phase loads. It is to be noted that a single-phase ac section can also be formed instead of the three-phase section, when the rating and number of the DERs and loads are small.

In Fig. 3a, the resistance R_f represents the switching and transformer losses, while the inductance L_f and capacitor C_f are connected to the output of the transformers to bypass the switching harmonics. A coupling inductance (L_T) is installed after C_f to provide the desired power sharing ratio among DERs [15].

The control of the ac section of the nanogrid consists of several sections, as discussed below:

2.1.1 Monitoring of SS Status

The nanogrid runs in off-grid or grid-connected status depending on the status of the SS. Isolation of the nanogrid from the utility feeder can be determined by observing the SS status and the availability of a proper data transmission infrastructure to send the data to the DERs. A suitable communication system for microgrids and nanogrids is presented in [26], and its impact on the stability of the system is discussed in [27].

Proper resynchronization should also be carried out before reconnection of the nanogrid to the utility feeder. This can be achieved by monitoring the magnitude and angle of the voltages in both sides of the SS via a phased-locked loop. Thereby, the SS will close only when the voltage magnitude and angles difference in its sides become smaller than a pre-defined minuscule limit. Likewise, a synchronization is also necessary for reconnection of a voltage-controlled DER to the nanogrid.

2.1.2 Control of Output Power

The converters of the DERs in the ac section operate in constant PQ mode, based on maximum power point tracking, when the nanogrid is in grid-connected status. However, they need to operate in voltage-controlled mode, based on droop, in off-grid status. To this end, dissimilar references are necessary for the converter output voltage:

When the nanogrid is in grid-connected status, the utility feeder forces the frequency and voltage in the nanogrid. Thus, the DERs can generate their rated power. Under such conditions, the active (p) and reactive power (q) at the coupling inductance terminal are [15]

$$
\begin{aligned}
p &= \frac{|V_T| \times |V_{cf}|}{\omega L_T} \sin(\delta_{cf} - \delta_T) \\
q &= \frac{|V_T|}{\omega L_T} \left(|V_{cf}| \cos(\delta_{cf} - \delta_T) - |V_T|\right)
\end{aligned}
\tag{1}
$$

where V_T is the voltage of the ac bus side of the coupling inductance, V_{cf} is the voltage across C_f and $V = |V| \angle\delta$ is the phasor representation of $v(t)$. The average active power (P) and reactive power (Q) are derived by passing p and q through a low-pass filter. Assuming the rated values for the output active and reactive powers and monitoring the voltage of the ac bus, the desired V_{cf} is calculated using (1). This is the voltage reference for the converters in grid-connected status, in Fig. 4.

On the other hand, the frequency and the output voltage of the converters of the DERs are modified according to the droop control and by the help of the pre-defined droop coefficients ($m > 0$, $n > 0$) in off-grid status. In this way, the DERs share the demand in the nanogrid with the specified ratio. The measured output powers of a converter are utilized to define the reference voltage at the converter output as [15]

$$\begin{aligned} \delta_{cf} &= \delta_{\text{ac, rated}} - m\left(P_{\text{ac}} - P_{\text{ac, rated}}\right) \\ \left|V_{cf}\right| &= V_{\text{ac, rated}} - n\left(Q_{\text{ac}} - Q_{\text{ac, rated}}\right) \end{aligned} \tag{2}$$

where the subscript rated shows the rated values and the subscript ac indicates the ac bus.

The block diagram of the control system for the converters of the DERs, connected to the ac bus, is shown in Fig. 4. Note that the SS status monitoring module issues a command to the reference selection block to select the proper reference.

It is to be highlighted that the frequency in the ac bus reduces by $\Delta\omega$ when a DER of this bus increases its output active power from zero to its rated value. Hence, the P–δ droop coefficient for these DERs is

$$m = \frac{\Delta\omega}{P_{\text{rated}}} \tag{3}$$

Assuming $\Delta\omega$ to be constant for all DERs with different ratings, the ratio of P–δ droop coefficient between any two DERs of the ac section of the nanogrid is

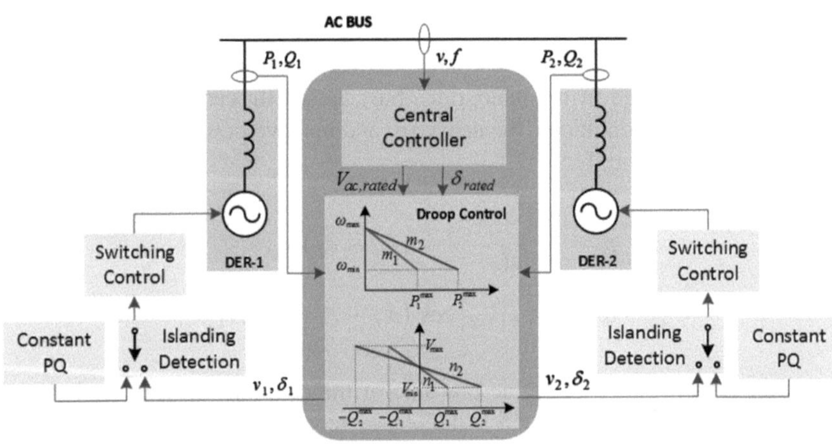

Fig. 4 Control block diagram of the DERs in the ac section of the nanogrid

$$\frac{m_1}{m_2} = \left(\frac{P_{\text{rated},1}}{P_{\text{rated},2}}\right)^{-1} \tag{4}$$

Similarly, the voltage magnitude in the ac bus reduces by ΔV, when a DER increases its output reactive power from zero to its rated value. Thus, the Q–V droop coefficient for these DERs is

$$n = \frac{\Delta V}{Q_{\text{rated}}} \tag{5}$$

Assuming ΔV to be constant for all DERs with different ratings, the ratio of Q–V droop coefficient between any two DERs of the ac section of the nanogrid is

$$\frac{n_1}{n_2} = \left(\frac{Q_{\text{rated},1}}{Q_{\text{rated},2}}\right)^{-1} \tag{6}$$

In [15], it was shown that the output active ratio and reactive power ratio among any two DERs are

$$\frac{P_1}{P_2} \approx \left(\frac{m_1}{m_2}\right)^{-1}, \quad \frac{Q_1}{Q_2} \approx \left(\frac{n_1}{n_2}\right)^{-1} \tag{7}$$

This technique is even suitable when the nanogrid has unbalanced and nonlinear loads, as discussed in [28]. Furthermore, it is also suitable when the nanogrid has an unequal distribution of single-phase DERs in the three-phase system of its ac section, as discussed in [29].

2.1.3 Switching Control of Converter

Converter switching control is in charge of turning on and off the IGBTs of the VSCs appropriately in such a way that the desired voltage is produced across capacitor C_f. This is discussed in detail in [30] and is not repeated here.

2.2 DC Bus

The dc section of the nanogrid is formed by another group of DERs and loads. Every DER is linked to the dc bus via a dc–dc converter. If the dc DER is a battery type, the dc–dc converter needs to be a bidirectional one to facilitate current flow in both directions, while for other DERs, a unidirectional converter is required. Boost converters are used for connecting the DERs to the dc bus in this chapter, as illustrated in Fig. 3b. The size of the inductor and capacitor filters in the dc–dc converter should be designed to limit the ripples of the current and voltage below acceptable limits. The control of the dc section of the nanogrid consists of several sections, as discussed below:

2.2.1 Monitoring of Tie-Converter Status

The islanding detection of the dc section of the nanogrid is realized by observing the tie-converter status and assuming the availability of a data transmission infrastructure to send these data to the DERs. It is to be noted that connection of the DERs to the dc bus does not need a synchronization technique.

2.2.2 Control of Output Power

The DERs of the dc bus may operate in two modes, based on the status of the tie-converter. When the tie-converter conducts power and the ac and dc buses are connected, the DERs of the dc bus are controlled in constant *P* mode, based on maximum power point tracking. Thus, the voltage of the dc bus is regulated by controlling the bidirectional power flow over the tie-converter. However, if the tie-converter is off and the buses are disconnected, the dc DERs operate under a droop-based voltage control mode. In such conditions, the output voltage of every DER converter is modified according to the droop control and the designed droop coefficients (m'). In this way, the DERs share the demand of the dc bus with the specified ratio. Hence, the measured output power of a converter (P) is utilized to define the voltage reference for converter output as [16, 17]

$$V_{\mathrm{dc}} = V_{\mathrm{dc,\,rated}} - m'\left(P_{\mathrm{dc}} - P_{\mathrm{dc,\,rated}}\right) \tag{8}$$

where the subscript *dc* denotes the dc bus.

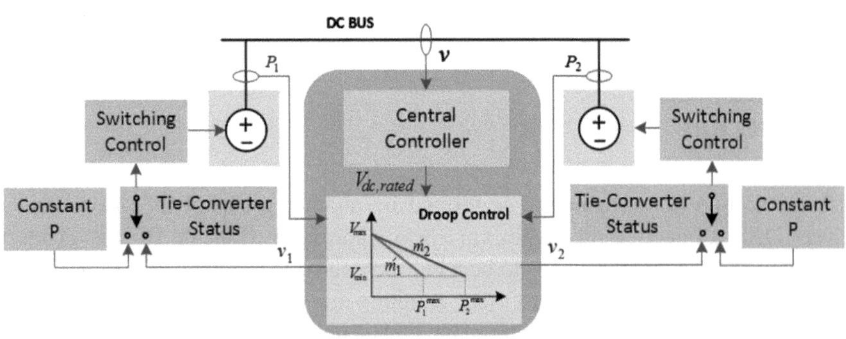

Fig. 5 Control block diagram of the DERs in the dc section of the nanogrid

The control block diagram of the converters of the DERs in the dc section of the nanogrid is shown in Fig. 5. It is to be noted that the tie-converter status monitoring module issues a command to the reference selection block to select the proper reference.

The voltage of the dc bus reduces by ΔV, when a DER of this bus increases its output active power from zero to its rated value. Hence, the *P–V* droop coefficient for these DERs is

$$m' = \frac{\Delta V}{P_{\text{rated}}} \tag{9}$$

Assuming ΔV to be constant for all DERs with different ratings, the ratio of *P–V* droop coefficient between any two DERs of the dc section of the nanogrid is

$$\frac{m'_1}{m'_2} = \left(\frac{P_{\text{rated},1}}{P_{\text{rated},2}}\right)^{-1} \tag{10}$$

Thus, it can be shown that the output active power ratio among any two DERs is [15]

$$\frac{P_1}{P_2} \approx \left(\frac{m'_1}{m'_2}\right)^{-1} \tag{11}$$

2.2.3 Switching Control of Converter

Converter switching control is in charge of turning on and off the IGBT in the boost converter (or the dc–dc converters in general) in such a way that the reference voltage is produced across the dc capacitor. This is discussed in detail in [31] and is not repeated here.

2.3 *Tie-Converter*

The tie-converter has a three-phase VSC, connected to the ac bus, through an LC filter, with a structure similar to those of DERs of the ac section of the nanogrid. In [18, 32], a current-controlled approach is used to control such converters. In this chapter, the voltage-controlled mode of the tie-converter is presented. In this technique, the tie-converter corrects the voltage of its point of common coupling (PCC) in the ac side to a balanced three-phase voltage. To this end, the converter has to exchange reactive power with the ac bus. Moreover, the tie-converter regulates the voltage of its dc side (i.e., the voltage of the dc bus). This is realized by regulating the voltage angle across C_f of the tie-converter ($\delta_{cf,\text{ TC, ref}}$) from [24]

$$\delta_{cf,\text{TC,ref}} = \left(k_P + \frac{k_I}{s}\right)\left(V_{\text{dc, desired}} - \widehat{V}_{\text{dc}}\right) \tag{12}$$

where $\widehat{V}_{\text{dc}}$ is the average of the dc bus voltage, $V_{\text{dc, desired}}$ = 350 V in this study, and k_P and k_I are PI coefficients. The tie-converter requires a switching control, similar to the switching control of the DERs of the ac bus.

3 Central Controller of the Nanogrid

The central controller of the nanogrid will regulate the frequency as well as the voltage of the ac bus along with the voltage of the dc bus. This is achieved by regulating the rated values in (2), (8) and (12), as discussed in [33]. This controller gets activated when the ac bus is isolated from the utility feeder. In the case that the frequency or the voltage of the ac bus or the voltage of the dc bus drops below or rise above acceptable pre-defined limits, the central controller takes action to shift down or up the rated values of the droop control. The new rated voltage and rated frequency level are calculated, based on the previous (old) rated levels and the dissimilarity between the actual voltage and frequency and the desired levels, from

$$\begin{aligned} \omega_{\text{ac, rated}}^{\text{new}} &= \omega_{\text{ac, rated}}^{\text{old}} + 2\pi(f_{\text{desired}} - f) \\ V_{\text{ac, rated}}^{\text{new}} &= V_{\text{ac, rated}}^{\text{old}} + V_{\text{ac, desired}} - |V_{\text{ac}}| \end{aligned} \tag{13}$$

where f and $|V_{\text{ac}}|$ are the frequency and voltage of the ac bus, respectively, while in this chapter, it is thought that $V_{\text{ac, desired}}$ = 240 V_{rms} and f_{desired} = 50 Hz. Likewise, the new rated voltage of the dc bus can be calculated from

$$V_{\text{dc, rated}}^{\text{new}} = V_{\text{dc, rated}}^{\text{old}} + V_{\text{dc, desired}} - \widehat{V}_{\text{dc}} \tag{14}$$

It is noteworthy that the central controller has a discrete monitoring time step with a slow time frame.

For the nanogrid system operating in Mode-1 and Mode-3, the central controller acts as the secondary control for every DER of both buses. In such a condition, the central controller defines the rated frequency and voltage for the DERs of the ac section and the rated voltage for the DERs of the dc section, as shown schematically in Fig. 6. In the case that the frequency and the voltage of the ac bus remain below the pre-defined minimum acceptable limits even after the operation of the central controller, load-shedding is required in this bus. When the nanogrid system is operating in Mode-2, the central controller is simply in charge of calculating the rated voltage of the DERs of the dc section. Likewise, if the voltage in the dc bus remains below the pre-defined minimum acceptable limits even after the operation of the central controller, load-shedding is required in this bus. The central controller is inactive when the nanogrid system operates in Mode-4.

4 Study Cases

The community loads are driven by a specific load profile. Furthermore, the DER generation is variable and intermittent in nature. Thereby, sudden changes in the loads, as well as generations, can be experienced in a nanogrid. Four modes of operation of the nanogrid are evaluated under some study cases here. Let us consider the network shown in Fig. 2 with 2 DERs in the ac bus (i.e., DER-1 and DER-2), 2 DERs in the dc bus (i.e., DER-3 and DER-4), and one load on each bus. The ratio of the nominal capacity of DER-1 and DER-2 is 1:2. Similarly, the ratio of the nominal capacity of DER-3 and DER-4 is 1:2. The studies are conducted based on the technical parameters provided in Table 1.

4.1 Evaluation of Mode-1

Let us assume that neither the tie-converter nor the SS is on. Hence, the ac bus, the dc bus, and the utility feeder are isolated from each other. In such a case, the DERs within both buses share the demand of that bus, and the central controller will control the voltages of the ac and dc bus in the range of ±10% and the frequency of the ac bus in the range of ±0.5 Hz. The nanogrid is thought to be at a steady-state condition initially, with a loading of 5 kW in the ac bus. At $t = 1$ s, the demand increases by 60% and decreases by 50% at $t = 2$ s. Figure 7a, b illustrates the output active power and reactive power of DER-1 and DER-2, linked to the ac bus. From this figure, it can be seen that the ratio among DER-1 and DER-2 is maintained as 1:2 during the study. The ac section demand changes are illustrated in Fig. 7c, for the same period. The frequency of the ac bus is illustrated in Fig. 7d and is within ±0.5 Hz boundaries. The voltage rms of the ac bus is shown in

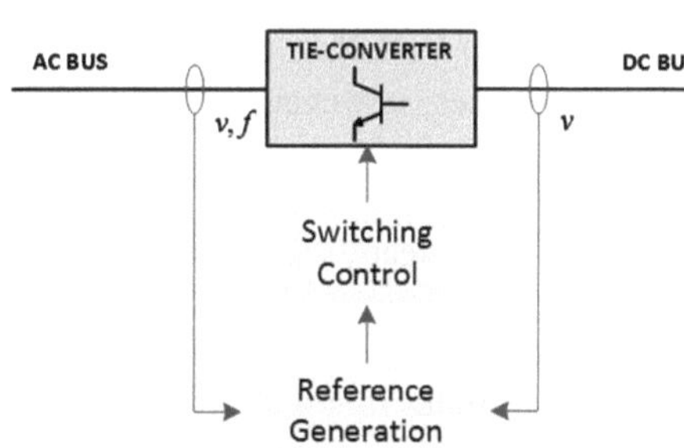

Fig. 6 The control block diagram of the tie-converter of the nanogrid

Table 1 Technical data of the nanogrid network under consideration

Utility feeder	11 kV rms L–L, 50 Hz, $R = 0.2\ \Omega$, $L = 10$ mH
Transformer	30 kVA, 11/0.415 kV, Dyn1, 50 Hz, $Z_I = 5\%$
AC bus	415 Vrms L–L
DC bus	350 V
DERs in AC bus	$V_{dc} = 350$ V, $a = 2$, $R_f = 0.1\ \Omega$, $L_f = 0.36$ mH, $C_f = 50\ \mu$F, $L_T = 2.27$ mH
	DER-1: $P_{1,\ rated} = 2$ kW, $L_T = 2.27$ mH, $m_1 = 3.1416$ rad/kW, $n_1 = 18$ V/kVAr DER-2: $P_{2,\ rated} = 4$ kW, $L_T = 1.135$ mH, $m_2 = 1.5708$ rad/kW, $n_2 = 9$ V/kVAr
DERs in DC bus	DER-3: $P_{3,\ rated} = 1$ kW, $m'_3 = 35$ V/kW, $V_{dc} = 350$ V
	DER-4: $P_{4,\ rated} = 2$ kW, $m'_4 = 17.5$ V/kW, $V_{dc} = 350$ V
Tie-converter (TC)	5 kVA, $V_{dc,\ ref} = 350$ V, $v_{cf,\ ref} = 240$ V, $a = 2$, $R_f = 0.1\ \Omega$, $L_f = 0.36$ mH, $C_f = 50\ \mu$F

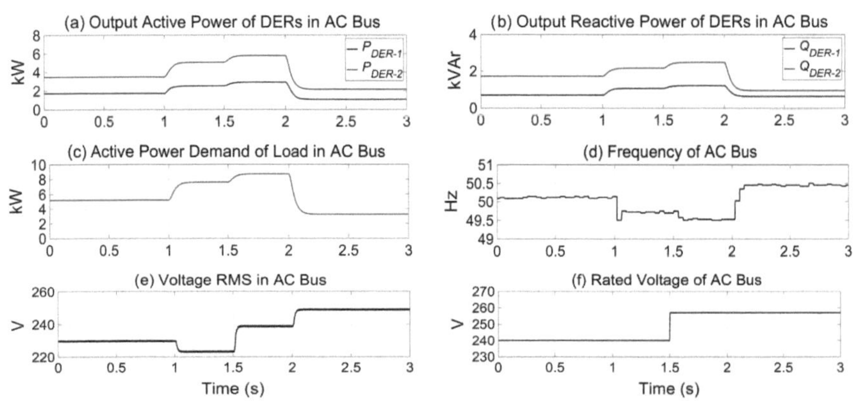

Fig. 7 Analysis outcome for the ac section of the nanogrid bus in Mode-1

Fig. 7e. From this figure, it can be seen that the voltage rms falls below −10% following the demand increase at $t = 1$ s. Assuming that the central controller has discrete time steps, it applies a modification in the droop curve of the DERs at $t = 1.5$ s. Hence, the rated voltage of the DERs is increased from 240 to 258 V, as shown in Fig. 7f. As a result of this change, the voltage magnitude of the ac bus restores to 240 V. Following the demand reduction at $t = 2$ s, the voltage rms of the ac bus increases; however, it is still below +10% limit. Thus, no further action is taken by the central controller.

The DERs of the dc bus supply the demand of this bus. It is assumed that the network is in a steady-state condition initially, with a loading of 3 kW in this bus. This demand increases by 100% at $t = 1$ s and decreases by 17% at $t = 2$ s. Figure 8a illustrates the output active power of DER-3 and DER-4, connected to this bus. From this figure, it can be seen that the ratio of the power of these DERs is sustained as 1:2. Figure 8b illustrates the demand of the dc bus. Figure 8c illustrates the voltage in the dc bus. From this figure, it can be seen that this voltage is retained within ±10% limits; thus, the central controller does not take any actions.

4.2 Evaluation of Mode-2

The tie-converter is thought to be off; thus, the ac bus and the dc bus are disconnected. However, the SS is supposed to be closed; thereby, the ac bus is coupled with the utility feeder. In this condition, the DERs of the ac bus function in constant PQ mode, whereas the DERs of the dc bus run in voltage droop control. The central controller is in charge of controlling the voltage of the dc bus only. Let us assume that the nanogrid is at a steady-state condition initially, with a loading of 7 kW in the ac bus. This demand increases by 29% at $t = 1$ s and decreases at $t = 2$ s by 66%. Figure 9a, b illustrates the output active power and reactive power of DER-1 and DER-2. From this figure, it can be seen that their output powers remain at the rated values at all times. Figure 9c, d illustrates the active and reactive power exchanged between the nanogrid and the utility feeder. From this figure, it can be seen that for $t < 2$ s, the utility feeder supplies a portion of the demand of the ac bus; however, for $t > 2$ s, the ac section of the nanogrid exports power into the utility feeder. The ac bus demand in this period is shown in Fig. 9e. Figure 9f illustrates the frequency of the ac bus. From this figure, it can be seen that the frequency is 50 Hz at all times. Figure 9e illustrates the voltage rms of the ac bus. From this figure, it can be seen that this voltage is not affected.

4.3 Evaluation of Mode-3

Now, let us assume that the tie-converter is on; thereby, the two buses are connected. The SS is thought to be off, and thereby, the nanogrid operates in off-grid

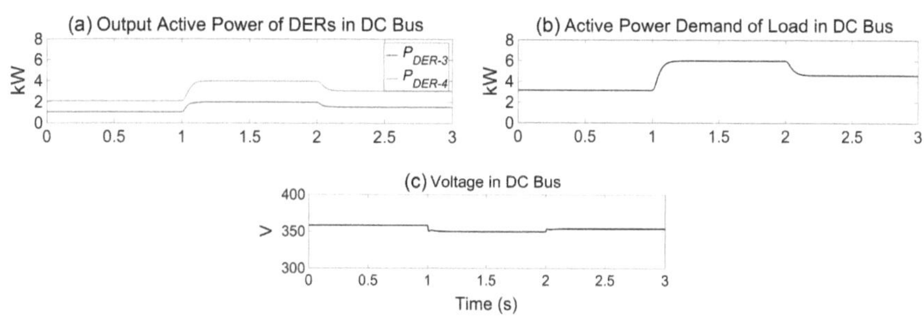

Fig. 8 Analysis outcome for the dc section of the nanogrid bus in Mode-1

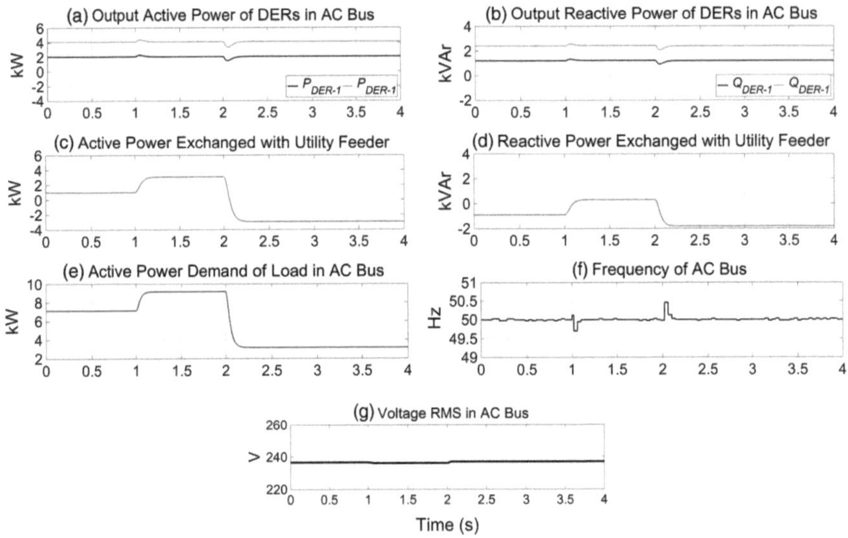

Fig. 9 Analysis outcome for the ac section of the nanogrid bus in Mode-2

mode. Under such a condition, the DERs of each bus supply the demand of that bus, while the tie-converter controls the voltage of both buses. In such a condition, the central controller is the backup of the tie-converter in voltage control. Several scenarios are considered and discussed below:

4.3.1 Demand Change in the AC Section

Let us assume that the nanogrid is at a steady-state condition initially, with a loading of 6 kW in the ac bus. This demand increases by 33% at t = 1 s, although the loading of the dc bus remains constant. Figure 10a illustrates the output active power of DER-1 and DER-2. From this figure, it can be seen that the demand variation is contributed by only these two DERs and the output active power ratio among them is maintained as 1:2. Figure 10b illustrates the output active power of DER-3 and DER-4. From this figure, it can be seen that the DERs of the dc bus do not contribute to this demand variation. The load variation is shown in Fig. 10c. The frequency and the voltage rms of the ac bus are shown, respectively, in Fig. 10d, e. From these figures, it can be seen that both frequency and voltage rms stay within acceptable limits. Figure 10f illustrates the voltage of the dc bus. From this figure, it can be seen that this voltage is not affected.

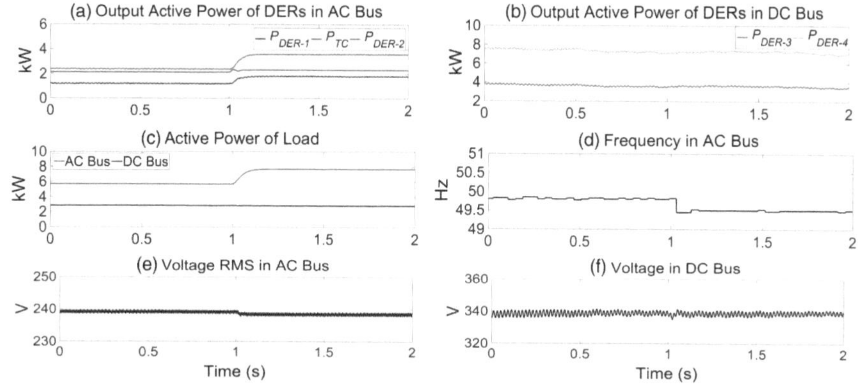

Fig. 10 Analysis outcome for the ac and dc sections of the nanogrid bus in Mode-3 (scenario-A)

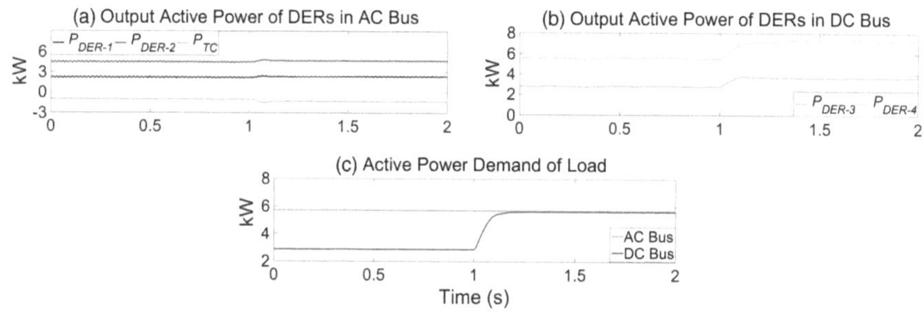

Fig. 11 Analysis outcome for the ac and dc sections of the nanogrid bus in Mode-3 (scenario-B)

4.3.2 Demand Change in the DC Section

Let us consider the nanogrid of scenario-A, with a loading of 3 kW in the dc bus. This demand increases by 100% at $t = 1$ s. Figure 11a illustrates the output active power of DER-1 and DER-2. From this figure, it can be seen that the output powers of the DERs of the ac bus are not affected by this demand variation. Thereby, the voltage rms and frequency of this bus will not be affected, either. Figure 11b illustrates the output active powers of DER-3 and DER-4. From this figure, it can be seen that these two DERs contribute to the demand variation, while the ratio of their output powers is maintained as 1:2. The demand variation is shown in Fig. 11c.

4.3.3 AC Bus DERs Operating in Maximum Capacity

Let us consider the nanogrid of scenario-A, where the DERs of the ac bus are thought to be operating in their maximum ratings. In this condition, the demand of the ac bus is approximately 6 kW where its 33% is supplied by the dc bus, through

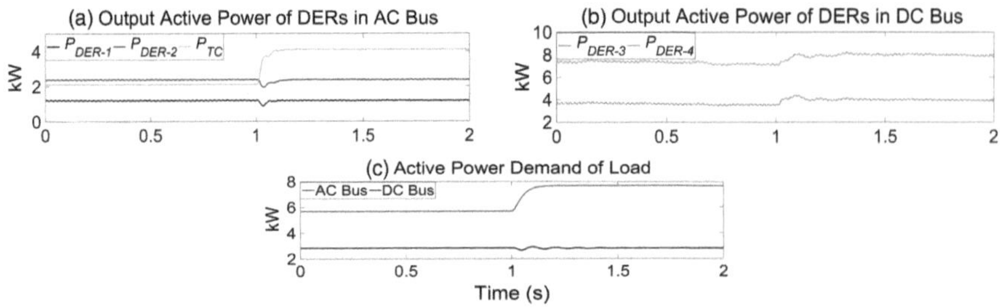

Fig. 12 Analysis outcome for the ac and dc sections of the nanogrid bus in Mode-3 (scenario-C)

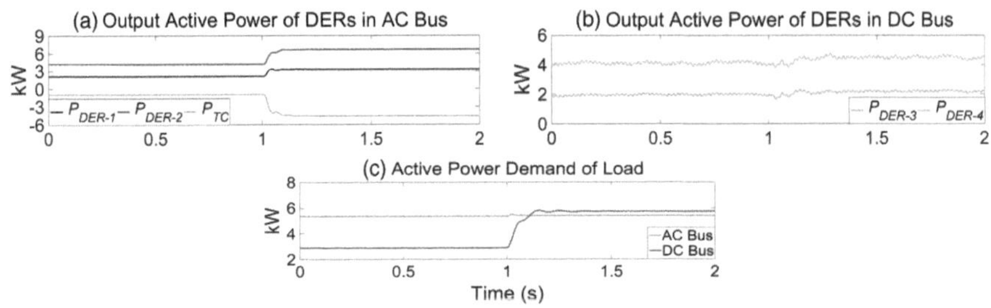

Fig. 13 Analysis outcome for the ac and dc sections of the nanogrid bus in Mode-3 (scenario-D)

the tie-converter. At $t = 1$ s, the demand of the ac bus increases by 30%. Figure 12a illustrates the output active power of the DER-1 and DER-2. From this figure, it can be seen that the output power of these DERs remains unaffected at their maximum capacities. Hence, the DERs in the dc bus contribute to this extra load and the power transfer over the tie-converter increases. Figure 12b illustrates the output active power of DER-3 and DER-4. From this figure, it can be seen that the output power of the DERs of the dc bus increases but is maintained as 1:2. The demand variations in the ac bus of the nanogrid are illustrated in Fig. 12c.

4.3.4 DC Bus DERs Operating in Maximum Ratings

Let us consider the nanogrid of scenario-B, where the DERs of the dc bus are operating in their maximum capacities. In this condition, the demand of the dc bus is 3 kW, while the DERs connected to this bus generate 6 kW. Therefore, the rest of the generated power is exported to the ac section through the tie-converter. Figure 13a illustrates the output active power of DER-1 and DER-2, as well as the active power, injected to the ac bus through the tie-converter. From this figure, it can be seen that only 1 kW is delivered to the ac bus through the tie-converter due to its power losses. At $t = 1$ s, the demand of the dc bus increases by 100%. Hence, the DERs of the ac bus contribute to this extra demand and the power transfer over the TC increases by 3 kW. As shown in Fig. 13a, the ratio of the output active power among the DER-1 and DER-2 is maintained as 1:2. Figure 13b illustrates the output active power of the DER-3 and DER-4. From this figure, it can be seen that the output power of these DERs remained as before. The demand variations in the nanogrid are illustrated in Fig. 13c.

4.4 *Evaluation of Mode-4*

Now, let us assume that both of the tie-converter and the SS are on; thereby, the buses are connected, and the nanogrid is operating in grid-connected mode. Under such a condition, the DERs of the ac bus function in constant PQ mode. Likewise, the DERs of the dc bus function in constant P mode. The extra demand of the nanogrid is supplied by the utility feeder, or the surplus power produced by the DERs is exported to the utility feeder. Let us assume that the nanogrid network is at a steady-state condition initially, with a loading of 4 kW in the ac bus and 1 kW in the dc bus. A load increase of 150% occurs in the ac bus at t = 1 s, while a load increase of 500% occurs in the dc bus at t = 2 s. Figure 14a, b illustrates the output active power of all the DERs, in both buses. From these figures, it can be seen that the ratio of the output powers is maintained as desired. For $t < 1$ s, the excess generated power by the DERs of the dc bus, minus the tie-converter losses, is delivered to the ac bus. Likewise, the excess power of the ac bus is exported to the utility feeder, as shown in Fig. 11c. As the demand of the ac bus increases at t = 1 s, the power transferred through the tie-converter remains unaffected. In such a condition, as the levels of the generated and consumed power in the ac bus are the same, the level of the exchanged power between the nanogrid and the utility feeder becomes zero. As the demand of the dc bus increases at t = 2 s, the direction of the transferred power via the tie-converter is inverted, and it takes 2 kW from the ac bus.

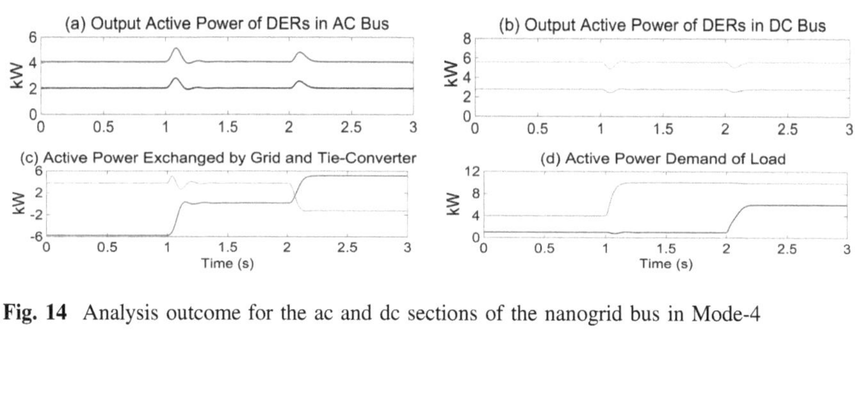

Fig. 14 Analysis outcome for the ac and dc sections of the nanogrid bus in Mode-4

5 Conclusion

This chapter presented the structure of a small-scale nanogrid system which facilitates the integration of renewable energy-based DERs in the electrical system of future community houses. This is an important technique for increasing the penetration of renewable energies for electricity generation and increases the sustainability of the electricity generation area. The nanogrid structure accommodates the ac and dc connection of the resources and supplies both dc- and ac-type loads within the community households. The two buses are interconnected via a tie-converter, which is in charge of controlling the voltage in both of them as it can facilitate a bidirectional power transfer between the buses. The chapter presented the primary control of each source as well as the central controller of the nanogrid. Four different operation modes are possible, based on the connection/disconnection status of the ac and the dc bus and the connection/disconnection status of the nanogrid with the utility feeder, as discussed in detail in the chapter.

References

1. L.A. de Souza Ribeiro, O.R. Saavedra, S.L. de Lima, and J. Gomes de Matos, "Isolated micro-grids with renewable hybrid generation: The case of Lençóis island," *IEEE Trans. on Sustainable Energy*, Vol. 2, No. 1, pp. 1–11, Jan. 2011.
2. M.V. Kirthiga, S.A. Daniel and S. Gurunathan, "A methodology for transforming an existing distribution network into a sustainable autonomous micro-Grid," *IEEE Trans. on Sustainable Energy*, Vol. 4, No. 1, pp. 31–41, Jan. 2013.
3. J. Bryan, R. Duke and S. Round, "Decentralized generator scheduling in a nanogrid using DC bus signaling," IEEE Power Engineering Society General Meeting, pp. 977–982, Vol. 1, June 2004.
4. R. P. S. Chandrasena, F. Shahnia, A. Ghosh and S. Rajakaruna, "Operation and control of a hybrid AC-DC nanogrid for future community houses," 24th Australasian Universities Power Engineering Conference (AUPEC), pp. 1–6, Australia, 2014.
5. A. Sannino, G. Postiglione and M.H.J. Bollen, "Feasibility of a DC network for commercial facilities," *IEEE Trans. on Industry Applications*, Vol. 39, No. 5, pp. 1499–1507, Sept./Oct. 2003.
6. N. Eghtedarpour and E. Farjah, "Distributed charge/discharge control of energy storages in a renewable-energy-based DC micro-grid," *IET Renewable Power Generation*, Vol. 8, No. 1, pp. 45–57, Jan. 2014.
7. N. Eghtedarpour and E. Farjah, "Power Control and Management in a Hybrid AC/DC Microgrid," *IEEE Transaction on Smart Grid*, Vol. 5, No. 3, pp. 1494–1505, May 2014.
8. C. Liang and M. Shahidehpour, "DC Microgrids: Economic Operation and Enhancement of Resilience by Hierarchical Control", *IEEE Transaction on Smart Grid*, Vol. 5, No. 5, pp. 2517–2526, Sept. 2014.
9. R.H. Lasseter, "Microgrids: distributed power generation," IEEE Power Engineering Society Winter Meeting, Vol. 1, pp. 146–149, 2001.
10. K. Jaehong, J.M. Guerrero, P. Rodriguez, et al. "Mode adaptive droop control with virtual output impedances for an inverter-based flexible ac microgrid," *IEEE Trans. on Power Electronics*, Vol. 26, No. 3, pp. 689–701, March 2011.

11. M. Savaghebi, A. Jalilian, J.C. Vasquez and J.M. Guerrero, "Secondary control scheme for voltage unbalance compensation in an islanded droop-controlled microgrid," *IEEE Trans. on Smart Grid*, Vol. 3, No. 2, pp. 797–807, June 2012.
12. F. Katiraei and M.R. Iravani, "Power management strategies for a microgrid with multiple distributed generation units," *IEEE Trans. on Power Systems*, Vol. 21, No. 4, pp. 1821–1831, Nov. 2006.
13. A. Mehrizi-Sani and R. Iravani, "Online set point modulation to enhance microgrid dynamic response: Theoretical foundation," *IEEE Trans. on Power Systems*, Vol. 27, No. 4, pp. 2167–2174, Nov. 2012.
14. M.B. Delghavi and A. Yazdani, "Islanded-mode control of electronically coupled distributed-resource units under unbalanced and nonlinear load conditions," *IEEE Trans. on Power Delivery*, Vol. 26, No. 2, pp. 661–673, April 2011.
15. F. Shahnia, R.P.S. Chandrasena, S. Rajakaruna and A. Ghosh, "Primary control level of parallel DER converters in system of multiple interconnected autonomous microgrids within self-healing networks," *IET Generation Trans. & Distribution*, Vol. 8, Issue 2, pp. 203–222, 2014.
16. X. Lu; J.M. Guerrero, S. Kai and J.C. Vasquez, "An improved droop control method for dc microgrids based on low bandwidth communication with dc bus voltage restoration and enhanced current sharing accuracy," *IEEE Trans. on Power Electronics*, Vol. 29, No. 4, pp. 1800–1812, April 2014.
17. T. Dragicevic, J.M. Guerrero, J.C. Vasquez and D. Skrlec, "Supervisory control of an adaptive-droop regulated dc microgrid with battery management capability," *IEEE Trans. on Power Electronics*, Vol. 29, No. 2, pp. 695–706, Feb. 2014.
18. P. C. Loh, Ding Li, Yi Kang Chai, and F. Blaabjerg, "Autonomous Operation of Hybrid Microgrid with AC and DC Subgrids," *IEEE Trans. on Power Electronics*, Vol. 28, No. 5, pp. 2214–2223, May 2013.
19. Xiaonan Lu, J. M. Guerrero, Kai Sun, J. C. Vasquez, R. Teodorescu, and Lipei Huang, "Hierarchical Control of Parallel AC-DC Converter Interfaces for Hybrid Microgrids," *IEEE Trans. on Smart Grid*, Vol. 5, No. 2, pp. 683–692, Mar. 2014.
20. Y. Ito, Y. Zhongqing and H. Akagi, "DC microgrid based distribution power generation system," IEEE 4th Int. Power Electronics and Motion Control Conf. (IPEMC), Vol. 3, pp. 1740–1745, Aug. 2004.
21. L. Xu and D. Chen, "Control and operation of a DC microgrid with variable generation and energy storage," *IEEE Trans. on Power Delivery*, Vol. 26, No. 4, pp. 2513–2522, Oct. 2011.
22. Z. Jiang and X. Yu, "Power electronics interfaces for hybrid DC and AC-linked microgrids," IEEE 6th Int. Power Electronics and Motion Control Conf. (IPEMC) pp. 730–736, May 2009.
23. P. Shanthi, U. Govindarajan, and D. Parvathyshankar, "Instantaneous power-based current control scheme for VAR compensation in hybrid AC/DC networks for smart grid applications," *IET Power Electronic.*, Vol. 7, No. 5, pp. 1216–1226, May 2014.
24. A. Ghosh and G. Ledwich, *Power Quality Enhancement using Custom Power Devices*, Kluwer Academic, 2002.
25. R. P. S. Chandrasena, F. Shahnia, S. Rajakaruna and A. Ghosh, "Dynamic operation and control of a hybrid nanogrid system for future community houses," *IET Generation, Transmission & Distribution*, vol. 9, no. 11, pp. 1168–1178, 2015.
26. M.A. Setiawan, F. Shahnia, R.P.S. Chandrasena and A. Ghosh, "Data communication network and its delay effect on the dynamic operation of distributed generation units in a microgrid," Power and Energy Engineering Conference (APPEEC), 2014 IEEE PES Asia-Pacific, Hong Kong, 2014, pp. 1–6.
27. M.A. Setiawan, F. Shahnia, S. Rajakaruna and A. Ghosh, "ZigBee-based communication system for data transfer within future microgrids," *IEEE Trans. on Smart Grid*, vol. 6, no. 5, pp. 2343–2355, 2015.
28. R.P.S. Chandrasena, F. Shahnia, S. Rajakaruna and A. Ghosh, "Control, operation and power sharing among parallel converter-interfaced DERs in a microgrid in the presence of

unbalanced and harmonic loads," 23rd Australasian Universities Power Engineering Conference (AUPEC), pp. 1–6, Australia, 2013.
29. F. Shahnia, and R.P.S. Chandrasena, "A three-phase community microgrid comprised of single-phase energy resources with an uneven scattering amongst phases," *International Journal of Electrical Power & Energy Systems*, Vol. 84, pp. 267–283, 2017.
30. F. Shahnia, S.M. Ami, and A. Ghosh, "Circulating the reverse flowing surplus power generated by single-phase DERs among the three phases of the distribution lines," *International Journal of Electrical Power & Energy Systems*, Vol. 76, pp. 90–106, March 2016.
31. N. Mohan, T.M. Undeland, and W.P. Robbins, *Power Electronics: Converters, Applications, and Design*, Wiley, 2002.
32. X. Liu, P. Wang and P.C. Loh, "A hybrid AC/DC microgrid and its coordination control," *IEEE Trans. on Smart Grid*, Vol. 2, No. 2, pp. 278–286, June 2011.
33. R.P.S. Chandrasena, F. Shahnia, A. Ghosh and S. Rajakaruna, "Secondary control in microgrids for dynamic power sharing and voltage/frequency adjustment," 24th Australasian Universities Power Engineering Conference (AUPEC), pp. 1–8, Australia, 2014.

Interconnected Microgrid Systems for Remote Areas

Farhad Shahnia

Abstract To reduce the frequency and necessity of load-shedding in a remote area microgrid during autonomous operation, islanded neighboring microgrids can be interconnected temporarily to support each other, if a proper overload management technique is in place and an extra generation capacity is available in the distributed energy resources in the neighboring microgrids. Otherwise, due to the unavailability of a utility feeder in remote areas, load-shedding is the only alternative to managing an overloaded microgrid. This way, the total demand of the system of coupled microgrids will be shared by all the distributed energy resources within these microgrids. To this end, a carefully designed overload management technique, protection systems, and communication infrastructure are required at the network and microgrid levels. In this chapter, first, the conditions and constraints based on which two neighboring microgrids are coupled are described. This is extended to develop an algorithm which identifies suitable microgrids to support the overloaded microgrid(s) when several neighboring microgrids exist in the distribution network.

Keywords Coupled microgrids · Overload management technique · Decision-making

1 Introduction

Supplying electricity to remote areas is a challenge due to the economic factors and the area accessibility [1, 2]. The electrical demand of remote/rural areas can be supplied by distributed energy resources (DERs) in an islanded scheme. Thus, their power systems can be reflected as a microgrid operating independently and isolated [3]. The design of these microgrids should be such that enough generation capacity is available in their local DERs to meet their demand [4–6]. Several independent

F. Shahnia (✉)
School of Engineering and Information Technology, Murdoch University, Perth, Australia
e-mail: f.shahnia@murdoch.edu.au

B. Azzopardi (ed.), *Sustainable Development in Energy Systems*,
DOI 10.1007/978-3-319-54808-1_3

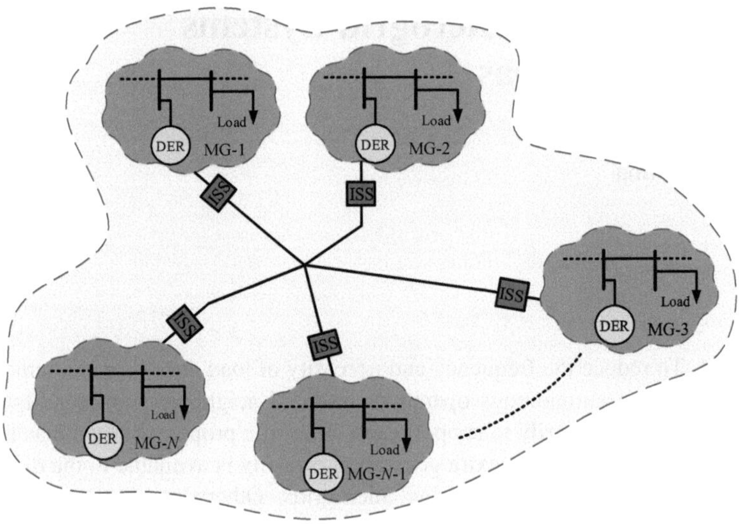

Fig. 1 A distribution network of a remote area that consists of a few isolated microgrids

microgrids can supply together a remote area/town. In such a case, each microgrid may have a different operator and each can be responsible for supplying the demand of a particular zone of the remote area. Figure 1 illustrates the distribution network of a remote area, composed of multiple microgrids.

The intermittency of solar and wind-type non-dispatchable DERs in addition to the uncertainties of the demand can lead to the power generation-demand mismatch (imbalance) in a microgrid. Any electricity generation deficiency or microgrid overloading will result in voltage and/or frequency drop. Several solutions can be considered when addressing the power imbalance problems in microgrids such as:

- load-shedding using under-frequency and under-voltage protective devices [7],
- using of energy storage systems such as batteries [8, 9],
- optimal capacity design of diesel/gas generator-based dispatchable DERs [10, 11],
- connecting the microgrid to a utility feeder [12],
- coupling the overloaded microgrid to one or a group of neighboring microgrid(s) [13].

Coupling of microgrids is introduced as a resolution to proliferate the DERs penetration in distribution networks [14]. Under such a scheme, each overloaded microgrid of Fig. 1 may be externally supported by one or a group of its neighboring microgrid(s). This can be realized by closing the normally open interconnecting static switches (ISSes), positioned between the microgrids. Without applying such a technique, load-shedding is the only option to maintain an overloaded remote area microgrid, unless any of the solutions mentioned above are already in place. To this end, the overload management technique (OMT) needs to

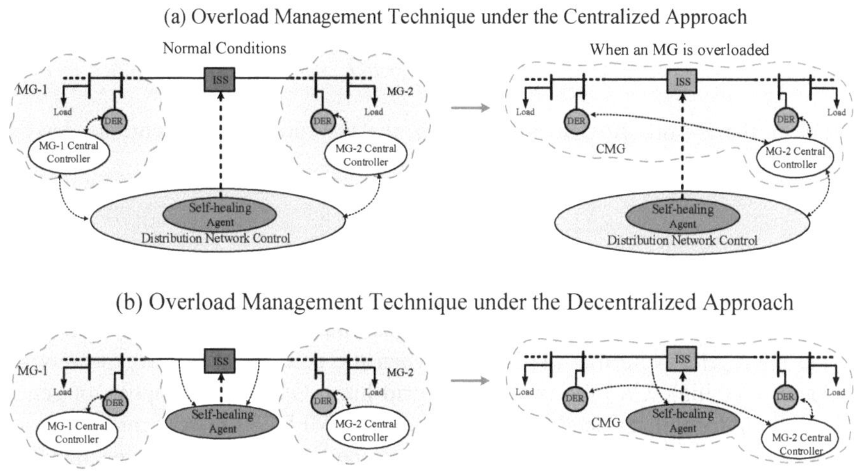

Fig. 2 Schematic illustration of the operation of the system of two microgrids in normal condition and after an overloading of one of them under the centralized and decentralized approach

facilitate the interconnection of the microgrids. The OMT may run under a centralized or a decentralized approach. The centralized approach is based on the assumption of the availability of a data communication system to send the generation amount details of the DERs instantaneously, while the decentralized approach is used when the data communication system is inaccessible momentarily or it does not exist. The OMT appears as a module of the distribution network controller under the centralized approach (see Fig. 2a) and as a module of the controller of the ISS, under the decentralized approach (see Fig. 2b).

The system of an islanded microgrid with an OMT offers more adaptability when overloaded. By coupling the microgrids with the help of the OMT, a portion of the microgrid demand will be supplied by the neighboring microgrid, and this will result in a reduced necessity and level of load-shedding.

2 Remote Area Network Under Consideration

First, let us consider the remote area network of Fig. 2 composed of MG-1 with N_1 DERs (i.e., DER_{11}, DER_{12}, …, DER_{1N1}), MG-2 with N_2 DERs (i.e., DER_{21}, DER_{22}, …, DER_{2N2}), and several loads in each microgrid. Each microgrid operates in islanded mode in normal conditions, and the voltage magnitude (V) and frequency (f) at the output of each DER are controlled based on its output active power (P) and reactive power (Q) with the help of a droop control as [15]

$$
\begin{aligned}
f &= f_{\max} - mP \\
V &= V_{\max} - nQ
\end{aligned}
\tag{1}
$$

where m, the P-f droop coefficients, and n, the Q-V droop coefficient, are derived from [15]

$$
\begin{aligned}
m &= (f_{\max} - f_{\min})/P_{\max} \\
n &= (V_{\max} - V_{\min})/Q_{\max}
\end{aligned}
\tag{2}
$$

while the *max* and *min* subscripts show the maximum and minimum allowed levels in the microgrid. Note that (1) is valid when the active and reactive powers are decoupled by high *X*/*R* ratio of the microgrid lines. Otherwise, adaptive or generalized droop techniques need to be deployed [16]. All DERs in both microgrids are assumed to have the same $f_{\max} - f_{\min}$ and the same $V_{\max} - V_{\min}$ values even though their capacities may be different.

3 Overload Management Technique (OMT)

Let us assume that MG-1 in Fig. 2 is overloaded, whereas MG-2 has surplus power. In such a case, the ISS between these microgrids must close so that the DERs of MG-2 supply a percentage of MG-1 demand. To this end, first, the overloaded microgrid should be detected, and then, an interconnection after which the system of coupled microgrid (CMG) may experience power mismatch must be prevented [13]. Once the condition and constraint for the interconnection are fulfilled, the OMT sends a command to the ISSes to close after proper synchronization. Figure 3 illustrates the operation flowchart of the OMT.

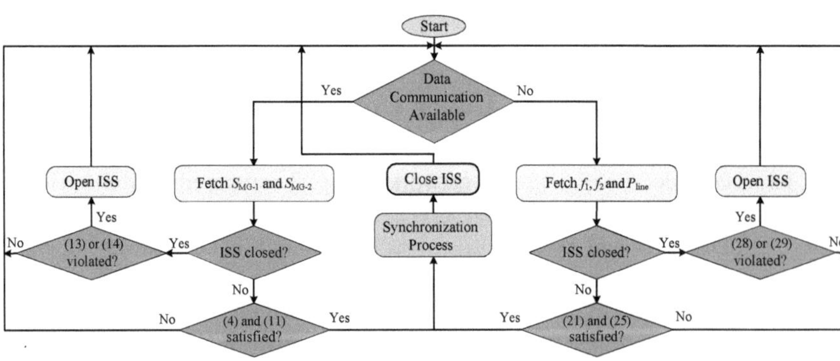

Fig. 3 Operation flowchart of the OMT

3.1 Centralized Approach

The unused power capacity (UPC) of MG-i ($\Delta S_{\text{MG-}i}$) can be defined as

$$\Delta S_{\text{MG-}i} = S_{\text{MG-}i}^{\text{cap}} - |S_{\text{MG-}i}| = \sum\nolimits_{j=1}^{N_i} S_{\text{DER-}ij}^{\text{cap}} - \left|\sum\nolimits_{j=1}^{N_i} S_{\text{DER-}ij}\right| \tag{3}$$

where $S_{\text{MG-}i}^{\text{cap}}$ and $|S_{\text{MG-}i}|$ are, respectively, the aggregated maximum capacities and the aggregated apparent powers of all DERs in MG-i and $i \in \{1,2\}$. Following an interconnection, all DERs will supply the total demand of the CMG. It is preferred to interconnect the microgrids if the UPC of MG-1 drops below $\alpha_1\, S_{\text{MG-1}}^{\text{cap}}$ where $0 < \alpha_1 < 1$ (e.g., $\alpha_1 = 0.1$), i.e., [13, 17]

$$\Delta S_{\text{MG-1}} < \alpha_1\, S_{\text{MG-1}}^{\text{cap}} \tag{4}$$

Nevertheless, the interconnection should be allowed if it can overload MG-2. Therefore, the interconnection constraint is to satisfy a minimum UPC in MG-2. Thus, the UPC of CMG should satisfy the prespecified criterion of

$$S_{\text{CMG}}^{\text{cap}} - |S_{\text{MG-1}} + S_{\text{MG-2}}| > \alpha_1\left(S_{\text{MG-1}}^{\text{cap}} + S_{\text{MG-2}}^{\text{cap}}\right) \tag{5}$$

From complex number rules, we have

$$|S_{\text{MG-1}} + S_{\text{MG-2}}| \le |S_{\text{MG-1}}| + |S_{\text{MG-2}}| \tag{6}$$

Considering (6), Eq. (5) is satisfied if

$$S_{\text{CMG}}^{\text{cap}} - (|S_{\text{MG-1}}| + |S_{\text{MG-2}}|) > \alpha_1\left(S_{\text{MG-1}}^{\text{cap}} + S_{\text{MG-2}}^{\text{cap}}\right) \tag{7}$$

Rewriting (7) as

$$\Delta S_{\text{MG-1}} + \Delta S_{\text{MG-2}} > \alpha_1\, S_{\text{MG-1}}^{\text{cap}} + \alpha_1 S_{\text{MG-2}}^{\text{cap}} \tag{8}$$

yields

$$\Delta S_{\text{MG-2}} > -\alpha_1 \psi\, S_{\text{MG-1}}^{\text{cap}} + \alpha_1\, S_{\text{MG-1}}^{\text{cap}} + \alpha_1 S_{\text{MG-2}}^{\text{cap}} \tag{9}$$

where

$$\psi = \Delta S_{\text{MG-1}} / \left(\alpha_1\, S_{\text{MG-1}}^{\text{cap}}\right) \tag{10}$$

Thus, the interconnection constraint becomes [13, 17]

$$\Delta S_{\text{MG-2}} > \alpha_1 \left[(1 - \psi)\, S_{\text{MG-1}}^{\text{cap}} + S_{\text{MG-2}}^{\text{cap}} \right] \tag{11}$$

If condition (4) and constraint (11) are fulfilled, the ISS controller receives a command from the distribution network controller to close. It is noteworthy that before the closing of the ISS, the stability of the new system should also be checked to avoid an interconnection which may cause instability. This is discussed in details in [18].

After the formation of a CMG, an apparent power of $S_{\text{Tie-line}}$ is supplied from MG-2 to MG-1. In such a condition, the ISS must open and isolate the microgrids when

$$S_{\text{MG-1}}^{\text{cap}} - \left| \sum_{j=1}^{N_1} S_{\text{DER-1}j} + S_{\text{Tie-line}} \right| > \alpha_1 \,(1 + \beta_1) S_{\text{MG-1}}^{\text{cap}} \tag{12}$$

The microgrids will remain interconnected while $S_{\text{Tie-line}}$ is above a limit, which can be defined from (12) considering (6) as [13, 17]

$$\left| S_{\text{Tie-line}} \right| > [1 - \alpha_1 (1 + \beta_1)] S_{\text{MG1}}^{\text{cap}} - \left| \sum_{j=1}^{N_1} S_{\text{DER-1}j} \right| \tag{13}$$

Also, the two microgrids should remain interconnected while the UPC of CMG fulfills [13, 17]

$$\Delta S_{\text{CMG}} > \alpha_1 \,(1 - \beta_2) S_{\text{CMG}}^{\text{cap}} \tag{14}$$

If either of (13) or (14) is violated, the ISS will open immediately to isolate the microgrids. β_1 and β_2 are considered in (13) and (14) to avoid continuous opening and closing (chattering) of the ISS.

3.2 Decentralized Approach

If the data communication system is momentarily inaccessible or if it does exists, the decentralized approach will be used in place of the centralized approach. In the decentralized approach, the microgrids are desired to interconnect when the unused active power capacity (UAPC) of MG-1 becomes lower than α_2 times of the total active power capacity of all the DERs of that microgrid $\left(P_{\text{MG-1}}^{\text{cap}} \right)$, i.e.,

$$P_{\text{MG-1}}^{\text{cap}} - \sum_{j=1}^{N_1} P_{\text{DER-1}j} < \alpha_2 \, P_{\text{MG-1}}^{\text{cap}} \tag{15}$$

Since all DERs have the same Δf (e.g., 1 Hz), it can be determined from (2) that (15) is fulfilled if the UAPC of each DER within MG-1 becomes lower than α_2 times of the capacity of that DER, i.e.,

$$P_{\text{DER-1}j}^{\text{cap}} - P_{\text{DER-1}j} < \alpha_2 P_{\text{DER-1}j}^{\text{cap}} \tag{16}$$

Henceforth, the criterion for the interconnection of these microgrids can be streamlined as

$$P_{\text{DER-1}j} > (1 - \alpha_2)\, P_{\text{DER-1}j}^{\text{cap}} \tag{17}$$

As a result, the interconnecting necessity of two microgrids can be concluded by only considering the loading of one DER in place of all the DERs.

From (1), the frequency of MG-1 is expressed as

$$f = f_{\max} - m_{\text{DER-1}j} P_{\text{DER-1}j} \tag{18}$$

Using (17) in (18), the frequency of MG-1 becomes

$$f < f_{\max} - m_{\text{DER-1}j}(1 - \alpha_2) P_{\text{DER-1}j}^{\text{cap}} \tag{19}$$

Substituting (2) in (19) produces

$$f < f_{\max} - \frac{f_{\max} - f_{\min}}{P_{\text{DER-1}j}^{\text{cap}}}(1 - \alpha_2) P_{\text{DER-1}j}^{\text{cap}} \tag{20}$$

Therefore, overloading of MG-1 can be clear if the frequency at MG-1 side of the ISS ($f_{\text{MG-1}}$) fulfills [13, 19]

$$f_{\text{MG-1}} < f_{\min} + \alpha_2 (f_{\max} - f_{\min}) \tag{21}$$

After interconnection, it is desired to have

$$\sum_{i=1}^{2} P_{\text{MG-}i}^{\text{cap}} - \sum_{i=1}^{2} \sum_{j=1}^{N_i} P_{\text{DER-}ij} > \alpha_2 \sum_{i=1}^{2} P_{\text{MG-}i}^{\text{cap}} \tag{22}$$

to avoid overloading of CMG. Using (18) in (22) gives

$$\sum_{i=1}^{2} P_{\text{MG-}i}^{\text{cap}} - \sum_{i=1}^{2} \sum_{j=1}^{N_i} (f_{\max} - f_{\text{MG-}i})/m_{ij} > \alpha_2 \sum_{i=1}^{2} P_{\text{MG-}i}^{\text{cap}} \tag{23}$$

which can be rewritten as

$$(1 - \alpha_2) \sum_{i=1}^{2} P_{\text{MG}i}^{\text{cap}} - \sum_{j=1}^{N_1} (f_{\max} - f_{\text{MG-1}})/m_{1j} > \sum_{j=1}^{N_2} (f_{\max} - f_{\text{MG-2}})/m_{2j} \tag{24}$$

Because $f_{\max} - f_{\text{MG-1}}$ is equal for all DERs of MG-1 and $f_{\max} - f_{\text{MG-2}}$ is equal for all DERs of MG-2, (24) can be expressed in the form of [13, 19]

$$f_{\text{MG}-2} > f_{\max} - \frac{(1-\alpha_2)\sum_{i=1}^{2} P_{\text{MG-}i}^{\text{cap}} - (f_{\max} - f_{\text{MG-1}})\sum_{j=1}^{N_1} 1/m_{1j}}{\sum_{j=1}^{N_2} 1/m_{2j}} \tag{25}$$

which is considered as the constraint for the interconnection. Therefore, two microgrids will be interconnected when (21) and (25) are both fulfilled.

After the interconnection of the microgrids, the CMG has a frequency of f_{CMG} and an active power of $P_{\text{Tie-line}}$ is supplied from MG-2 to MG-1. Thus, the power consumed in MG-1 (P_{MG1}) becomes

$$P_{\text{MG1}} = \sum_{j=1}^{N_1} P_{\text{DER-1}j} + P_{\text{Tie-line}} \tag{26}$$

The CMG must be divided into two isolated microgrids if $P_{\text{Tie-line}}$ becomes less than a limit. Substituting (26) in (15) gives

$$P_{\text{MG1}}^{\text{cap}} - \left(\sum_{j=1}^{N_1} (f_{\max} - f_{\text{CMG}})/m_{1j} + P_{\text{Tie-line}}\right) < \alpha_2(1+\beta_3)P_{\text{MG1}}^{\text{cap}} \tag{27}$$

which can be rewritten as [13, 19]

$$P_{\text{Tie-line}} > [1 - \alpha_2(1+\beta_3)]P_{\text{MG1}}^{\text{cap}} - (f_{\max} - f_{\text{CMG}})\sum_{j=1}^{N_1} 1/m_{1j} \tag{28}$$

The ISS must remain closed if f_{CMG} is above a limit of [13, 19]

$$f_{\text{CMG}} > f_{\min} + \alpha_2(1-\beta_4)(f_{\max} - f_{\min}) \tag{29}$$

which is defined using (21). If either of (28) or (29) is not satisfied, the ISS will open immediately to isolate the microgrids. β_3 and β_4 are also used in (28) and (29) to avoid chattering in the ISS operation.

The UPC levels of an isolated microgrid and a CMG are a function of α_1, β_1, and β_2, as shown in Fig. 4 [13].

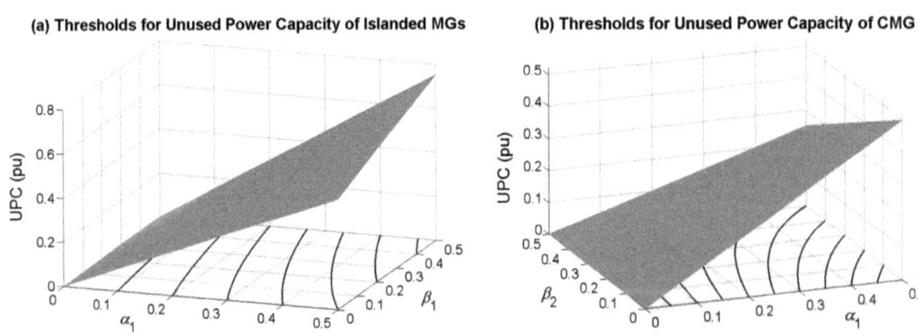

Fig. 4 UPC levels of two microgrids in isolated mode and after forming a CMG with respect to the OMT coefficients

4 Synchronization of Microgrids Before Their Coupling

Synchronization is necessary before interconnecting two microgrids as each microgrid may have a different voltage and frequency. Synchronization starts as soon as ISS controller receives a close command from the OMT. The ISS will close when the difference of voltage angles and magnitudes on across the ISS become very small (i.e., smaller than a predefined limit) [15]. In this way, significant distortions in the voltage, current, and power waveforms are avoided in the microgrids. The duration of synchronization is a function of the difference of the voltage angles at two sides of the ISS ($\Delta\theta$) and the difference between the frequencies of the microgrids (Δf) at the time of the command. For example, the synchronization duration is shown as a function of $\Delta\theta$ for $0.25 \leq \Delta f \leq 1$ Hz in Fig. 5. This figure illustrates that for each Δf, the synchronization duration increases linearly with the increase in $\Delta\theta$. It also illustrates that for each $\Delta\theta$, the synchronization duration increases with the decrease in Δf [13].

5 Required Level of Load-Shedding

A portion of the demand of the overloaded microgrid is supplied by the neighboring microgrid, with the help of the OMT [13]. In the network of Fig. 2, if the microgrids are islanded, the level of load-shedding in MG-1 $\left(S_{\text{MG}-1}^{\text{shed}}\right)$ can be calculated using (4) as

$$S_{\text{MG-1}}^{\text{shed}} = |S_{\text{MG-1}}| - (1 - \alpha_1) S_{\text{MG-1}}^{\text{cap}} \tag{30}$$

However, when the microgrids are coupled, the level of load-shedding does not only depends on the instantaneous power generated by the DERs of MG-1 and their capacities but is also a function of the instantaneous power generated by the DERs

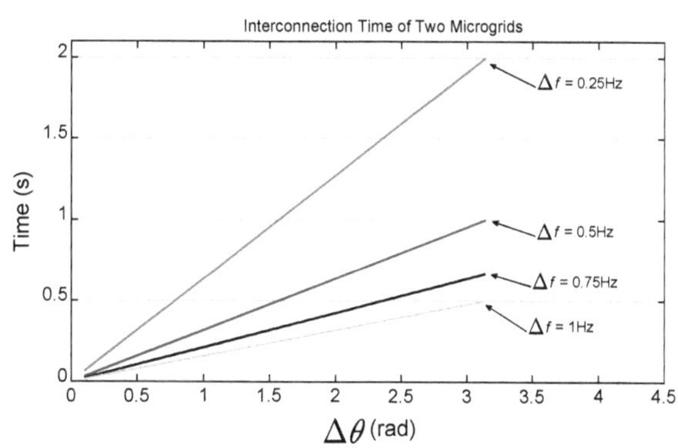

Fig. 5 Synchronization time for the interconnection of two microgrids for several $\Delta\theta$ and Δf values

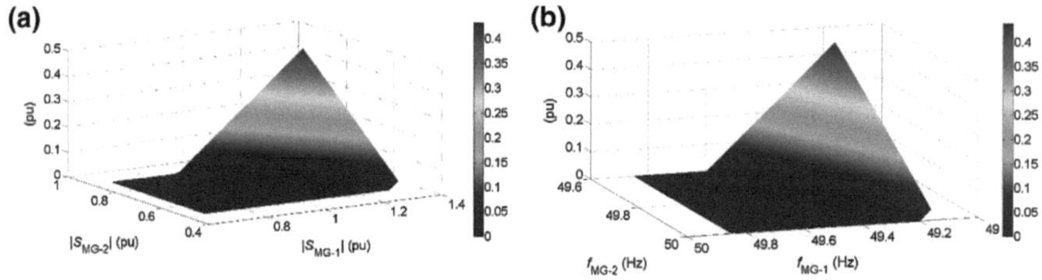

Fig. 6 Load-shedding amount in MG-1 in interconnected mode under **a** centralized approach, **b** decentralized approach

of MG-2 and their capacities. Therefore, $S^{\text{shed}}_{\text{MG-1}}$ can be defined for a CMG, using (7), as

$$S^{\text{shed}}_{\text{MG-1}} = \sum\nolimits_{i=1}^{2} |S_{\text{MG-}i}| - (1 - \alpha_1) \sum\nolimits_{i=1}^{2} S^{\text{cap}}_{\text{MG-}i} \tag{31}$$

Under the decentralized approach, $P^{\text{shed}}_{\text{MG}-1}$ can be derived from (21) as

$$P^{\text{shed}}_{\text{MG-1}} = \sum\nolimits_{j=1}^{N_1} (f_{\max} - f_{\text{MG-1}})/m_{1j} - (1 - \alpha_2) P^{\text{cap}}_{\text{MG-1}} \tag{32}$$

while if a CMG is formed, $P^{\text{shed}}_{\text{MG}-1}$ can be defined from (23) as

$$P^{\text{shed}}_{\text{MG-1}} = \sum\nolimits_{i=1}^{2} \sum\nolimits_{j=1}^{N_i} (f_{\max} - f_{\text{MG-}i})/m_{ij} - (1 - \alpha_2) \sum\nolimits_{i=1}^{2} P^{\text{cap}}_{\text{MG-}i} \tag{33}$$

As an example, Fig. 6 shows the level of load-shedding in MG-1 when coupled to MG-2 under the centralized and decentralized approaches for a given system [13]. Figure 6a shows that load-shedding is not necessary for MG-1 when MG-2 is loaded below 70% prior to the interconnection. It also indicates that MG-1 a load-shedding of 5 and 10% is required if MG-2 is loaded by 75 and 80%, respectively. Similarly, Fig. 6b shows that load-shedding is not necessary for MG-1 when MG-2 frequency is higher than 49.8 Hz prior to the interconnection. It also illustrates that a load-shedding of 5 and 10% is required in MG-1 if MG-2 frequency falls to 49.75 and 49.7 Hz, respectively.

6 Network Dynamic Performance with the OMT

Let us consider the network of Fig. 2, where each microgrid has two DERs. Three scenarios are taken into account for the OMT operating under the centralized approach, and three scenarios when it is running under the decentralized approach. Table 1 lists the events considered in each simulation case in time sequence for the centralized approach, while Table 2 is the list of the events for the decentralized

Table 1 Events in the studies of Fig. 7

t (s)	Event	Consequence
Case-1		
0	The microgrids are isolated and operating at steady state	
1	MG-1 load increases by 83%	S_{MG-1} increases to supply the load
1.2	Interconnection condition (4) and constraint (11) are fulfilled	The OMT sends the ISS closing command
2.16	Synchronization of microgrids is accomplished	ISS closes to form a CMG
3.2	MG-2 load increases by 60%	S_{CMG} increases to supply the load
3.5	Conditions (13) and (14) are not violated	No action is taken
Case-2		
0	The microgrids are isolated and operating at steady state	
1	MG-1 load increases by 32%	S_{MG-1} increases to supply the load
1.2	Interconnection condition (4) and constraint (11) are fulfilled	The OMT sends the ISS closing command
1.75	Synchronization of microgrids is accomplished	ISS closes to form a CMG
3.3	MG-1 load reduces by 44%	S_{CMG} decreases to supply the load
3.5	Condition (13) is violated	ISS opens, and microgrids are isolated
Case-3		
0	The microgrids are isolated and operating at steady state	
1	MG-1 load increases by 102%	S_{MG-1} increases to supply the load
1.2	Interconnection condition (4) and constraint (11) are fulfilled	The OMT sends the ISS closing command
2.95	Synchronization of microgrids is accomplished	ISS closes to form a CMG
3.5	MG-1 load increases by 5%	S_{CMG} increases to supply the load
3.6	Condition (14) is violated	ISS opens, and microgrids are isolated

approach. In each scenario, MG-1 demand increases, and it becomes overloaded. The OMT detects that MG-1 is overloaded and sends a closing command to the controller of the ISS. The ISS closes following synchronization between the microgrids. Later, extra demand variations are considered following which the OMT determines whether the ISS should open or remain closed. Figure 7 illustrates the dynamic simulation results of each microgrid for the centralized approach, while Fig. 8 shows the results for the decentralized approach [13].

Table 2 Events in the studies of Fig. 8

t (s)	Event	Consequence
Case-1		
0	The microgrids are isolated and operating at steady state	
1	MG-1 load increases by 115%	P_{MG-1} increases to supply the load
1.2	Interconnection condition (21) and constraint (25) are fulfilled	The OMT sends the ISS closing command
2.49	Synchronization between the microgrids is accomplished	ISS closes to form a CMG.
3.3	MG-2 load increases by 40%	P_{CMG} increases to supply the load
3.5	Conditions (28) and (29) are not violated	No action is taken
Case-2		
0	The microgrids are isolated and operating at steady state	
1	MG-1 load increases by 86%	P_{MG-1} increases to supply the load
1.2	Interconnection condition (21) and constraint (25) are fulfilled	The OMT sends the ISS closing command
2.39	Synchronization of microgrids is accomplished	ISS closes to form a CMG
3.3	MG-1 load reduces by 54%	P_{CMG} decreases to supply the load
3.5	Condition (28) is violated	ISS opens, and microgrids are isolated
Case-3		
0	The microgrids are isolated and operating at steady state	
1	MG-1 load increases by 94%	P_{MG-1} increases to supply the load
1.2	Interconnection condition (21) and constraint (25) are fulfilled	The OMT sends the ISS closing command
2.28	Synchronization of microgrids is accomplished	ISS closes to form a CMG
3.3	MG-2 load increases by 150%	P_{CMG} increases to supply the load
3.5	Condition (29) is violated	ISS opens, and microgrids are isolated

7 Complexity of Coupling Microgrids

Now, let us assume the distribution network of Fig. 1 composed of N isolated microgrids where N' microgrids among them are overloaded. In general, the alternatives that can be considered to support the overloaded microgrids are combinations of single microgrids, two microgrids, three microgrids…, and $N - N'$ microgrids among $N - N'$ microgrids. Therefore, the possible alternatives number is [20]

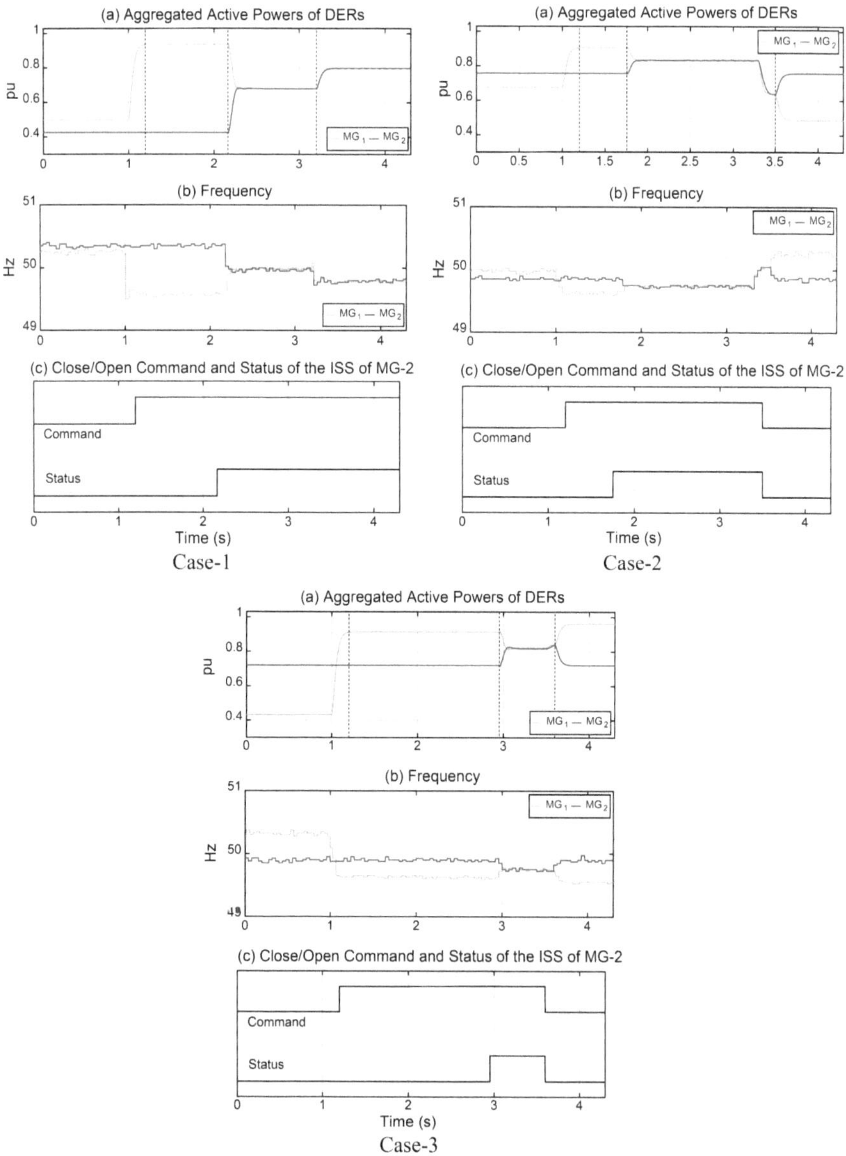

Fig. 7 Dynamic performance of the network of Fig. 2 under the OMT operating in the centralized approach

$$N_A = C_1^{N\text{-}N'} + C_2^{N\text{-}N'} + C_3^{N\text{-}N'} + \ldots + C_{N\text{-}N'}^{N\text{-}N'} = 2^{N\text{-}N'} - 1 \tag{34}$$

where $C_a^b = b!/[(b–a)!\times a!]$ and $a! = a \times (a - 1) \times \ldots \times 1$. As an example, for a distribution network with $N = 5$ microgrids in which MG-1 is overloaded (i.e.,

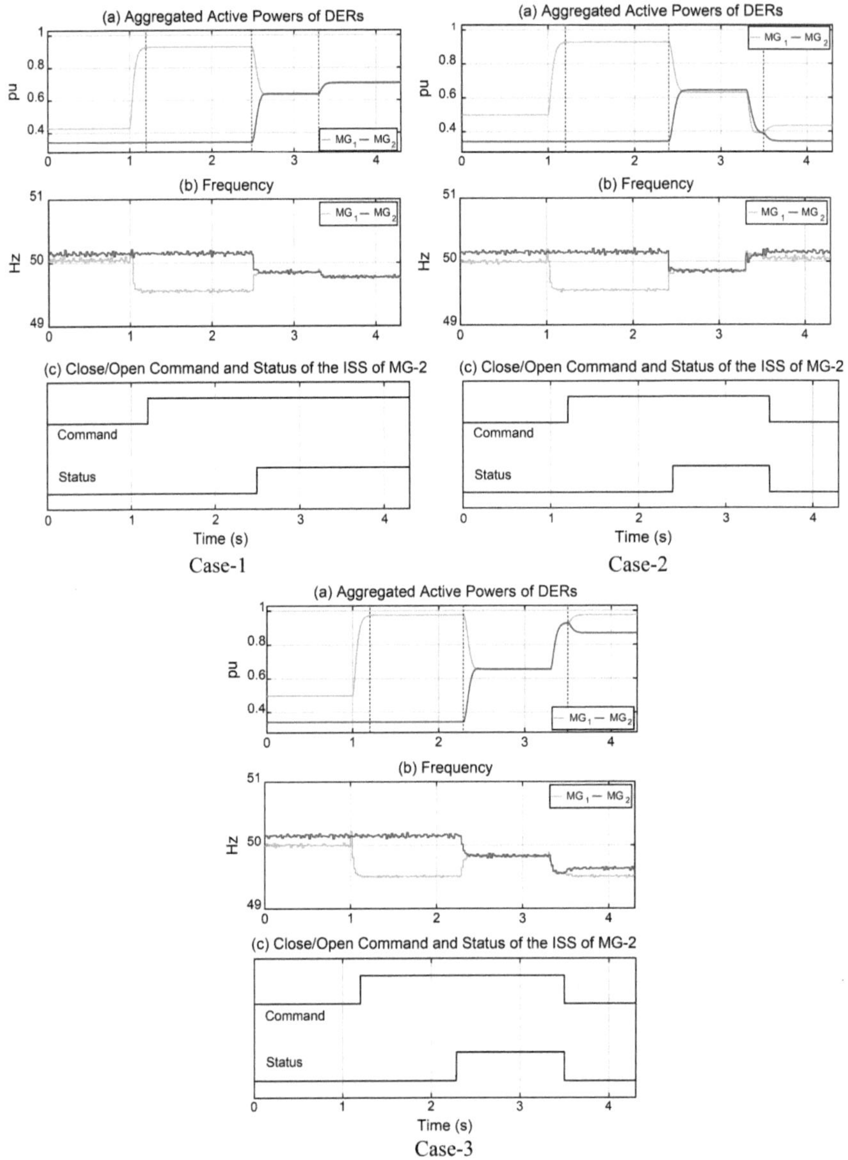

Fig. 8 Dynamic performance of the network of Fig. 2 under the OMT operating in the decentralized approach

$N' = 1$), the alternative microgrids are [{2}, {3}, {4}, {5}, {2,3}, {2,4}, {2,5}, {3,4}, {3,5}, {4,5}, {2,3,4}, {2,3,5}, {2,4,5}, {3,4,5}, {2,3,4,5}]. Thereby, selection of the appropriate microgrid(s) to couple with the overloaded microgrid among the possible alternatives is a difficult task because of the large number of possible alternatives, several criteria for the selection, and the different weightings for each criterion.

To select the appropriate alternative, the OMT can utilize a decision-making algorithm. To this end, the OMT, situated as a module within the distribution network controller, has to send/receive information with the central controllers of the microgrid and the ISSes continuously, as illustrated schematically in Fig. 9. It will receive the power generated by the dispatchable DERs ($P_{\text{disp-DER (MG-}i)}$) and the power consumed by the loads ($P_{\text{load (MG-}i)}$). At the first step, the OMT will recognize the microgrid(s) that are overloaded. Then, it will take a suitable action according to the conditions of the network (e.g., the number of overloaded microgrids and the available UPC in the other microgrids). It computes the UPC of MG-i from

$$UPC_i = \sum P_{\text{disp-DER (MG-}i)} - \sum P_{\text{load (MG-}i)} 4 \forall\, i \in \{1, \ldots, N_{\text{A}}\} \tag{35}$$

It is preferred to keep the UPC of every microgrid above the limit of [20]

$$UPC_i \geq \alpha_1 \sum P^{\max}_{\text{disp-DER (MG-}i)} \tag{36}$$

where $\sum P^{\max}_{\text{disp-DER (MG-}i)}$ is the overall capacity of the dispatchable DERs in MG-i, and $0 < \alpha_1 < 1$ (e.g., $\alpha_1 = 0.1$) is a safety margin. (36) is considered as the overloading condition of a microgrid. If condition (36) is valid for all microgrids of the distribution network, the OMT does not need to take any actions. But if it is invalid for one or some of the microgrids, the OMT assesses the availability of enough surplus power in the network from

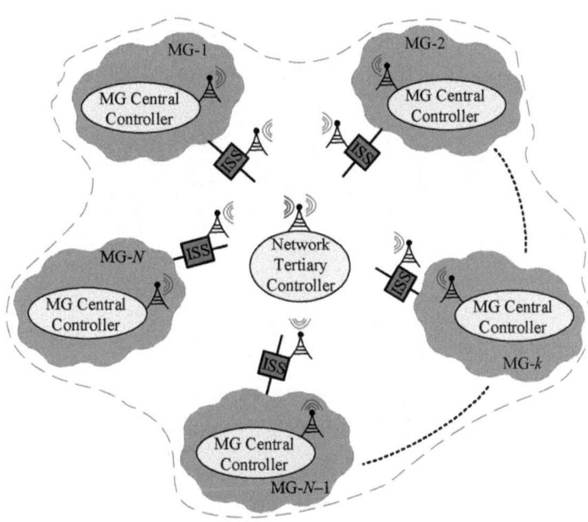

Fig. 9 Schematic diagram of the communication links for the OMT

$$\sum_{i=1}^{N} UPC_i \geq \alpha_1 \sum_{i=1}^{N} \sum P_{\text{disp-DER (MG-}i)}^{\max} \tag{37}$$

Equation (37) is assumed as the constraint for coupling the microgrids. If condition (36) shows that one or some of the microgrids are overloaded, and constraint (37) shows that the distribution network has enough surplus power to support the overloaded microgrids, the OMT continues to select an appropriate alternative.

If condition (36) is unacceptable for $n_1 = N - 1$ microgrids but constraint (37) is valid, the OMT concludes that the only alternative is coupling all microgrids. It runs a power flow analysis to confirm that coupling all microgrids does not result in nonstandard deviations in the voltage and frequency of the CMG. The interconnection is allowed if the maximum variation of voltage in all buses (ΔV) and the maximum variation of frequency (ΔF) satisfy

$$\Delta V \leq 0.1 \text{ and } \Delta F \leq 0.5 \tag{38}$$

If condition (36) is unacceptable for $1 < n_1 < N$ microgrids, the OMT concludes that several alternatives are possible, and decision-making is necessary. Then, it runs the decision-making function to frame all possible alternatives and to select the appropriate. After the selection of an appropriate alternative, then it sends a closing command to the relevant ISS(es). Yet, if no appropriate alternative is chosen, load-shedding is inevitable.

8 Multi-criteria Decision-Making Algorithm

Decision-making is a technique to choose the favorable alternative from a set of alternatives (A), based on N_c criteria (c), where each criterion may have a dissimilar weighting (w) where $\sum_j w_j = 1$. The multi-criteria decision-making problem can be formulated as a matrix in the form of [21]

$$\begin{array}{c} \\ A_1 \\ A_2 \\ \vdots \\ A_{N_A} \end{array} \begin{array}{c} \begin{array}{cccc} c_1(w_1) & c_2(w_2) & \cdots & c_{N_C}(w_{N_C}) \end{array} \\ \begin{bmatrix} x_{11} & x_{12} & \cdots & x_{1N_C} \\ x_{21} & x_{22} & \cdots & x_{2N_C} \\ \vdots & \vdots & \ddots & \vdots \\ x_{N_A1} & x_{N_A2} & \cdots & x_{N_AN_C} \end{bmatrix} \end{array} \tag{39}$$

where x_{uv} is the performance of alternative A_u from the criterion c_v point of view. Equation (39) will be altered to take into account the weightings of the criteria as

$$\begin{array}{c} \\ A_1 \\ A_2 \\ \vdots \\ A_{N_A} \end{array} \begin{array}{c} \begin{array}{cccc} c_1 & c_2 & \cdots & c_{N_C} \end{array} \\ \begin{bmatrix} w_1 x_{11} & w_2 x_{12} & \cdots & w_{N_C} x_{1N_C} \\ w_1 x_{21} & w_2 x_{22} & \cdots & w_{N_C} x_{2N_C} \\ \vdots & \vdots & \ddots & \vdots \\ w_1 x_{N_A 1} & w_2 x_{N_A 2} & \cdots & w_{N_C} x_{N_A N_C} \end{bmatrix} \end{array} = \begin{bmatrix} X_1 \\ X_2 \\ \vdots \\ X_{N_A} \end{bmatrix} \quad (40)$$

The evaluation results for the alternatives (X) are calculated as [22–24]

$$\begin{aligned} X_u &= \text{average}(w_v x_{uv}) = \sum\nolimits_{u=1}^{N_C} w_v x_{uv} / N_C \\ X_u &= \text{product}(w_v x_{uv}) = \prod\nolimits_{u=1}^{N_C} w_v x_{uv} \\ X_u &= \min(w_v x_{uv})_{v=1,\ldots,N_C} \\ X_u &= \alpha_3 . \min(w_v x_{uv}) + (1 - \alpha_3) . \max(w_v x_{uv}) \end{aligned} \quad (41)$$

where max(·) and min(·) and are the maximum and minimum functions, respectively, and $\alpha_3 \in [0,1]$ (e.g., $\alpha_3 = 0.75$) is the optimist coefficient. The alternative that has the highest X will have the highest priority and will be chosen by the decision-making algorithm.

The four aggregators of (41) may not essentially choose a similar alternative. To address this issue, the risk index (R) will be computed for each selected alternative in the matrix form of [21, 22]

$$\begin{array}{c} \\ A_1 \\ A_2 \\ \vdots \\ A_{N_A} \end{array} \begin{array}{c} \begin{array}{cccc} c_1(x_1^{\max}) & c_2(x_2^{\max}) & \cdots & c_2(x_2^{\max}) \end{array} \\ \begin{bmatrix} r_{11} & r_{12} & \cdots & r_{1N_M} \\ r_{21} & r_{22} & \cdots & r_{2N_M} \\ \vdots & \vdots & \ddots & \vdots \\ r_{N_A 1} & r_{N_A 2} & \cdots & r_{N_A N_M} \end{bmatrix} \end{array} = \begin{bmatrix} R_1 \\ R_2 \\ \vdots \\ R_{N_A} \end{bmatrix} \quad (42)$$

where r_{uv} for criterion c_v is the variation of the performance of each alternative from the alternative with the highest performance. It is computed from

$$r_{uv} = w_v x_{uv} - x_v^{\max} \quad (43)$$

where $x_v^{\max}$ is the maximum of $w_v x_{uv}$ for criterion c_v among all possible alternatives. From (42), the risk index (R_u) for alternative A_u is defined as the maximum of r_{uv}. If the aggregators of (41) choose dissimilar alternatives, the chosen alternative with a lower risk index will be selected by the decision-making algorithm as the appropriate alternative.

The outcome of the decision-making algorithm rests on the considered weightings for the criteria. Thereby, these weightings need to be defined cautiously.

In complex systems like power systems, there are no systematic ways to define such weightings; however, an acceptable technique is a survey of the experts of the field. A group of experts can contribute to the selection of the weightings based on their experiences. Eventually, the weighting of each criterion will be computed as the average of the weightings nominated by the experts for that specific criterion.

The flowchart of the multi-criteria dynamic decision-making algorithm is illustrated in Fig. 10.

First, the decision-making algorithm evaluates the qualification of the alternatives based on the four criteria of [25]

- Criterion-1: Consent/veto of a microgrid to couple with an overloaded neighboring microgrid,
- Criterion-2: Availability of surplus power in microgrids,
- Criterion-3: Maximum variation in the voltage, and
- Criterion-4: Maximum variation in the frequency.

And the performance of alternative A_u for these criteria is computed from

$$x_{u1} = \begin{cases} 0 & \text{MG vetoes coupling} \\ 1 & \text{MG consents coupling} \end{cases}$$
$$x_{u2} = \begin{cases} 1 & \kappa_u \geq 3 \\ 0.5\,(\kappa_u - 1) & 1 \leq \kappa_u < 3 \\ 0 & \kappa_u < 1 \quad \text{or} \quad x_{u1} = 0 \end{cases}$$
$$x_{u3} = \begin{cases} -10\Delta V_u + 1 & 0 \leq \Delta V_u \leq 0.1 \\ 10\,\Delta V_u + 1 & -0.1 \leq \Delta V_u < 0 \\ 0 & 0.1 < |\Delta V_u| \quad \text{or} \quad x_{u1}.x_{u2} = 0 \end{cases} \tag{44}$$
$$x_{u4} = \begin{cases} -2\Delta F_u + 1 & 0 \leq \Delta F_u \leq 0.5 \\ 2\,\Delta F_u + 1 & -0.5 \leq \Delta F_u < 0 \\ 0 & 0.5 < |\Delta F_u| \quad \text{or} \quad x_{u1}.x_{u2}.x_{u3} = 0 \end{cases}$$

where κ_u is the ratio of the UPC in microgrid(s) of alternative A_u versus the power deficiency level (PDL) of overloaded MG-j (i.e., κ_u = UPC$_u$/PDL$_{\text{MG-}j}$). If any of the above four criteria disqualify an alternative, the performance of that alternative for the remaining criteria is neglected, i.e.,

$$x_{uv} = \begin{cases} 0 & \text{if } x_{u1}.x_{u2}.x_{u3}.x_{u4} = 0 \\ x_{uv} & \text{if } x_{u1}.x_{u2}.x_{u3}.x_{u4} \neq 0 \end{cases} \quad \forall v \in \{5, \ldots, N_C\} \tag{45}$$

Another six criteria are also utilized, as discussed below [25]:

- Criterion-5: Loss in the interconnecting lines between the microgrids

The power loss rests on the distance between the overloaded microgrid and the selected microgrid(s) and their line impedances. It is calculated from the power flow

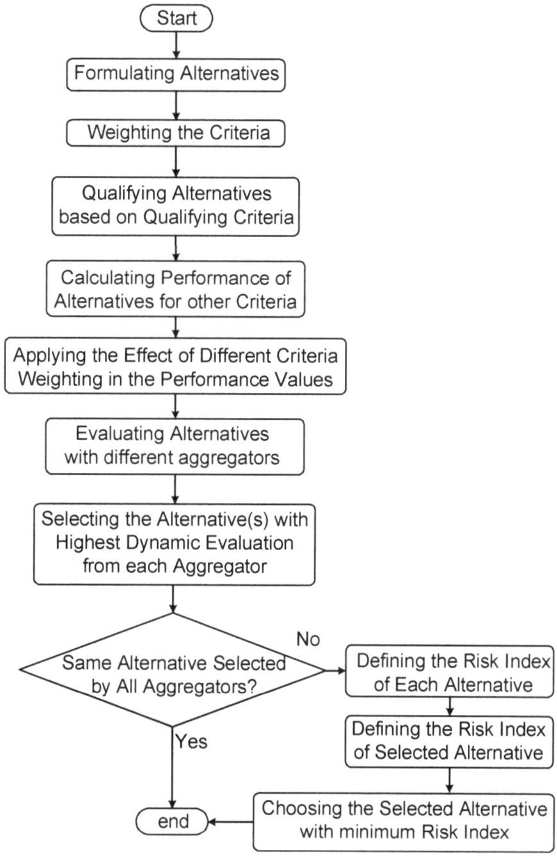

Fig. 10 Flowchart of the multi-criteria decision-making algorithm

analysis for each alternative. It is then normalized in the [0, 1] range in which 0 represents the alternative with the maximum loss, and 1 represents the alternative with the minimum loss. The performance of each alternative is computed for this criterion from

$$x_{u5} = 1 - \frac{P_{\text{loss},u}}{\max(P_{\text{loss},u})_{u=1,\ldots,N_A}} \tag{46}$$

- Criterion-6: Electricity price

A microgrid operator may sell electricity to neighboring microgrids with a dissimilar price in regard to its local costumers. Also, the price may change dynamically over time. Assuming E_i as the electricity unit price of MG-i (in $/kWh), the equivalent electricity cost to be paid by the operator of the overloaded microgrid to the operators of microgrid(s) in alternative A_u can be computed from

$$E_{\text{cost},u} = \sum\nolimits_{i=1}^{p} \beta_{i,u} E_i \tag{47}$$

From (47), the performance of alternative A_u is computed for this criterion from

$$x_{u6} = 1 - \frac{E_{\text{cost},u}}{\max\left(E_{\text{cost},u}\right)_{u=1,\ldots,N_A}} \tag{48}$$

- Criterion-7 and 8: Reliability

It is preferred for the overloaded microgrid to couple with microgrid(s) with higher reliability to reduce the probability of interruption to the demand of the overloaded microgrid, after an interconnection. Several reliability indices may be utilized such as system average interruption frequency index (SAIFI), momentary average interruption frequency index (MAIFI), and system average interruption duration index (SAIDI). RB_u^{f} and RB_u^{d} indices are defined for alternative A_u composed of p microgrids as

$$\begin{aligned} RB_u^{\text{f}} &= \sum\nolimits_{i=1}^{p} \lambda_1 . SAIFI_{i,u} + \lambda_2 . MAIFI_{i,u} \\ RB_u^{\text{d}} &= \sum\nolimits_{i=1}^{p} SAIDI_{i,u} \end{aligned} \tag{49}$$

where $\{\lambda_1, \lambda_2\} \in [0, 1]$ and $\lambda_1 + \lambda_2 = 1$ are the assumed weightings for SAIFI and MAIFI, respectively. From (49), the performance of alternative A_u is calculated for these criteria as

$$\begin{aligned} x_{u7} &= 1 - \frac{RB_u^{\text{f}}}{\max\left(RB_u^{\text{f}}\right)_{u=1,\ldots,N_A}} \\ x_{u8} &= 1 - \frac{RB_u^{\text{d}}}{\max\left(RB_u^{\text{d}}\right)_{u=1,\ldots,N_A}} \end{aligned} \tag{50}$$

- Criterion-9: Supply security

A microgrid, which has a larger generated power from its non-dispatchable DERs, may have a higher UPC in its dispatchable DERs. When such a microgrid is coupled with an overloaded microgrid, the overloaded microgrid does not have a high supply security as any unpredicted reduction in the generated power of the non-dispatchable DERs may result in the CMG overloading. Supply security index (SS_u) is defined for alternative A_u composed of p microgrids as

$$SS_u = \frac{\sum_{i=1}^{p} \sum P_{\text{disp-DER(MG-}i)}}{\sum_{i=1}^{p} \sum P_{\text{load(MG-}i)}} \tag{51}$$

where $\sum P_{\text{load(MG-}i)}$ represents the average active power consumption of all loads in MG-i. From (51), the performance of alternative A_u can be computed for this criterion from

$$x_{u9} = 1 - \frac{SS_u}{\max{(SS_u)}_{u=1,\ldots,N_A}} \tag{52}$$

- Criterion-10: CO_2 emissions

It is preferred to choose an alternative with fewer CO_2 emissions. Assuming the total CO_2 emission for MG-i as Em_i, CO_2 emission index is defined for alternative A_u composed of p microgrids as

$$\text{CO}_{2u} = \sum_{i=1}^{p} \beta_{i,u} Em_i \tag{53}$$

From (53), the performance of alternative A_u is computed for this criterion from

$$x_{u10} = 1 - \frac{\text{CO}_{2u}}{\max{(\text{CO}_{2u})}_{u=1,\ldots,N_A}} \tag{54}$$

9 System Performance with the Decision-Making Technique

Let us assume the network of Fig. 1 with three isolated microgrids, each constituted of non-dispatchable solar and wind-based DERs, a diesel generator, and some essential and nonessential loads [25]. In Example-1 and 2, one overloaded microgrid is defined, and the distribution network is found to have enough surplus power to support the overloaded microgrid. Table 3 provides the assumed data for all three microgrids and the computed UPC and PDL for Example-1. This table shows that MG-1 is overloaded while MG-2 and MG-3 have surplus power. Thus, three alternatives of A_1 = {MG-2}, A_2 = {MG-3}, and A_3 = {MG-2, MG-3} are available. Table 4 lists the assumed data for each microgrid of the alternatives for this example. These data are employed in decision-making algorithm to define the performance of each alternative for each criterion, as seen from Table 5. Table 6 illustrates the decision-making matrix of (39). Assuming the normalized weightings for the criteria as listed in Table 7, the weighted decision-making matrix is computed from (40) and is given in Table 8. It is then normalized, as provided in

Table 3 Assumed data for the considered microgrids and their calculated UPC and PDL levels [kW]

	$\sum P_{\text{load}}$	$\sum P_{\text{disp-DER}}^{\max}$	$\sum P_{\text{disp-DER}}$	P_{wind}	P_{pv}	UPC	PDL	β
MG-1	63	58	61.4424	1.3985	0.1591	–	9.2424	–
MG-2	55	63	37.4820	17	0.5180	25.52	–	0.562
MG-3	34	49	16.8372	17	0.1628	32.16	–	0.437

Table 4 Assumed data for each microgrid

	E	SAIFI	MAIFI	SAIDI	Em
MG-2	0.2748	5.1883	6.2373	56.4762	3.5261
MG-3	0.3158	4.1663	11.2216	43.4203	4.4508

Table 5 Comparison of the parameters of possible alternatives to couple with overloaded MG-1

	Consent	κ_u	$\lvert\Delta V_u\rvert$	$\lvert\Delta F_u\rvert$	$P_{\text{loss},u}$	$E_{\text{cost},u}$	RB_u^{f}	RB_u^{d}	SS_u	$CO_{2,u}$
A_1	1	2.76	0.080	0.383	6	0.275	11	56	0.681	3.526
A_2	1	3.48	0.071	0.136	5	0.316	15	43	0.495	4.451
A_3	1	6.24	0.014	0.151	5.562	0.293	27	99	0.610	3.931

Table 6 Calculated decision-making matrix

	c_1	c_2	c_3	c_4	c_5	c_6	c_7	c_8	c_9	c_{10}
A_1	1	0.880	0.197	0.234	0	0.130	0.706	0.668	0.681	0.208
A_2	1	1	0.294	0.728	0.166	0	0.604	0.744	0.495	0
A_3	1	1	0.858	0.699	0.073	0.073	0.311	0.412	0.610	0.117

Table 7 Assumed weightings for the criteria in decision-making algorithm

w_1	w_2	w_3	w_4	w_5	w_6	w_7	w_8	w_9	w_{10}
0.133	0.120	0.120	0.120	0.093	0.106	0.080	0.080	0.080	0.006

Table 9. This table is then used to define the evaluation result of each alternative, using the aggregators of (41), as given in Table 10. For Example-1, MG-1 is overloaded, and alternative A_3 is the preferred choice of all aggregators and is selected to couple to MG-1. The data of Table 5 should be examined to find out the reason for the selection of A_3 over the two other alternatives. This table shows that A_3 has the finest performance from criterions 2 and 3 points of view, while it has an average performance from criterions 4, 5, 6, 9, and 10 points of view. Also, A_3 has the poorest performance from criterions 7 and 8 points of view. Finally, the decision-making algorithm selects this alternative since criterions 2, 3, 4, and 6 had higher weightings against criterions 7 and 8.

Table 8 Calculated weighted decision-making matrix

	c_1	c_2	c_3	c_4	c_5	c_6	c_7	c_8	c_9	c_{10}
A_1	0.133	0.106	0.024	0.028	0	0.014	0.056	0.053	0.054	0.014
A_2	0.133	0.120	0.035	0.087	0.015	0	0.049	0.060	0.040	0
A_3	0.133	0.120	0.103	0.084	0.007	0.008	0.025	0.033	0.049	0.008

Table 9 Normalized weighted decision-making matrix

	c_1	c_2	c_3	c_4	c_5	c_6	c_7	c_8	c_9	c_{10}
A_1	0.333	0.306	0.146	0.414	0	0.640	0.435	0.366	0.381	0.640
A_2	0.333	0.347	0.218	0.438	0.696	0	0.373	0.408	0.278	0
A_3	0.333	0.347	0.636	0.421	0.304	0.360	0.192	0.226	0.341	0.360

Table 10 Evaluation results from considered four aggregators and their chosen alternative

Alternatives	Average	Min	Product	Hurwitz
A_1	3.3890	0	0	0.1600
A_2	3.0903	0	0	0.1739
A_3	3.5207	0.1917	0.0000	0.3027
Selected Alternative	A_3	A_3	A_3	A_3

Table 11 Normalized weighted decision-making matrix

	c_1	c_2	c_3	c_4	c_5	c_6	c_7	c_8	c_9	c_{10}
A_1	0.333	0.333	0.304	0.343	0	0.381	0.428	0.400	0.571	0.633
A_2	0.333	0.333	0.114	0.348	0.243	0.279	0.419	0.412	0.069	0
A_3	0.333	0.333	0.581	0.308	0.756	0.338	0.151	0.187	0.358	0.366

Table 12 Evaluation results from considered four aggregators and their chosen alternative along with their risk index

Alternatives	Average	Min	Product	Hurwitz
A_1	0.373	0	0	0.158
A_2	0.255	0	0	0.104
A_3	0.371	0.151	0.000	0.302
Selected Alternative	A_1	A_3	A_3	A_3
Risk Index	0.756	0.277	0.277	0.277
Preferred Alternative	A_3			

Example-2 illustrates another case in which MG-3 is overloaded. The normalized weighted decision-making matrix for this case is given in Table 11, while the results of the evaluation by aggregators of (41) for each alternative are listed in Table 12. This table illustrates that alternatives A_1 and A_3 are chosen by the

Table 13 Risk matrix

	c_1	c_2	c_3	c_4	c_5	c_6	c_7	c_8	c_9	c_{10}
A_1	0	0	0.277	0.005	0.756	0	0	0.012	0	0
A_3	0	0	0	0.040	0	0.043	0.277	0.225	0.213	0.267

aggregators. Thereby, a risk matrix is computed for these alternatives from (42) to (43) and is given in Table 13. From this table, alternative A_3 is eventually chosen as the appropriate alternative as it has a smaller risk index.

10 Conclusion

An overload management technique, discussed in this chapter, can be used to decrease the load-shedding rate of an overloaded remote area microgrid, by interconnecting it with appropriate neighboring microgrid(s). The OMT can operate under a centralized and a decentralized approach, based on the availability of the communication system. Diverse interconnection and isolation conditions are utilized for each approach. Simulation studies illustrate that the OMT can define the overloaded microgrid and the appropriate neighboring microgrid after evaluating the interconnection conditions. The OMT can also divide the CMG into isolated microgrids if coupling conditions or requirements are disturbed. A multi-criteria decision-making algorithm is used when numerous neighboring microgrids are available. To this end, the possible alternatives are first formulated, qualified, and assessed based on some predefined criteria and the most suitable one is selected.

References

1. R. Paleta, A. Pina and C.A.S. Silva, "Polygeneration energy container: designing and testing energy services for remote developing communities," *IEEE Trans. on Sustainable Energy*, Vol. 5, pp. 1348–1355, 2014.
2. M. Arriaga, C.A Canizares and M. Kazerani, "Northern lights: access to electricity in Canada's northern and remote communities," *IEEE Power and Energy Magazine*, Vol. 12, pp. 50–59, 2014.
3. K. Ubilla, *et al.* "Smart microgrids as a solution for rural electrification: ensuring long-term sustainability through cadastre and business models" *IEEE Trans. on Sustainable Energy*, Vol. 5, pp. 1310–1318, 2014.
4. Y.M. Atwa, E.F.El-Saadany, M.M.A. Salama, *et al.*, "Adequacy evaluation of distribution system including wind/solar DG during different modes of operation," *IEEE Trans. on Power System*, vol. 26, no. 4, pp. 1945–1952, 2011.
5. J.M. Guerrero, J.C. Vasquez, J. Matas, *et al.*, "Hierarchical control of droop-controlled ac and dc microgrids- A general approach toward standardization," *IEEE Trans. on Industrial Electronics*, vol. 58, no. 1, pp. 158–172, 2011.
6. B. Kroposki, R. Lasseter, T. Ise, *et al.*, "Making microgrids work," *IEEE Power and Energy Magazine*, vol. 6, no. 3, pp. 40–53, 2008.

7. Y.Y. Hong, M.C. Hsiao, Y.R. Chang, *et al.*, "Multiscenario underfrequency load shedding in a microgrid consisting of intermittent renewables," *IEEE Trans. on Power Delivery*, vol. 28, no. 3, pp. 1610–1617, 2013.
8. E. Pashajavid, F. Shahnia and A. Ghosh, "Overload management of autonomous microgrids," IEEE 11th International Conference on Power Electronics and Drive Systems (PEDS), pp. 73–78, Australia, 2015.
9. S. Bahramirad, W. Reder and A. Khodaei, "Reliability-constrained optimal sizing of energy storage system in a microgrid," *IEEE Trans. on Smart Grid*, vol. 3, no. 4, pp. 2056–2062, 2012.
10. S.A. Arefifar and Y.A.R.I. Mohamed, "DG Mix, reactive sources and energy storage units for optimizing microgrid reliability and supply security," *IEEE Trans. on Smart Grid*, vol. 5, no. 4, pp. 1835–1844, 2014.
11. S.A. Arefifar, Y.A.I. Mohamed and T.H.M. EL-Fouly, "Optimum microgrid design for enhancing reliability and supply-security," *IEEE Trans. on Smart Grid*, vol. 4, no. 3, pp. 1567–1575, 2013.
12. R. Majumder, A. Ghosh, G. Ledwich and F. Zare, "Power management and power flow control with back-to-back converters in a utility connected microgrid," *IEEE Trans. on Power Systems*, vol. 25, no. 2, pp. 821–834, 2010.
13. E. Pashajavid, F. Shahnia, and A. Ghosh, "Development of a self-healing strategy to enhance the overloading resilience of islanded microgrids," *IEEE Trans. Smart Grid*, vol. 8, no. 2, pp. 868–880, 2017.
14. R.H. Lasseter, "Smart distribution: coupled microgrids," *Proceedings of the IEEE*, vol. 99, no. 6, pp. 1074–1082, 2011.
15. F. Shahnia, R.P.S. Chandrasena, S. Rajakaruna, A. Ghosh, "Primary control level of parallel distributed energy resources converters in system of multiple interconnected autonomous microgrids within self-healing networks" *IET Generation Transmission Distribution*, vol. 8, pp. 203–222, 2014.
16. M. Fathi and H. Bevrani, "Statistical cooperative power dispatching in interconnected microgrids," *IEEE Trans. on Sustainable Energy.*, vol. 4, no. 3, pp. 586–593, 2013.
17. E. Pashajavid, F. Shahnia and A. Ghosh, "Overloading conditions management in remote networks by coupling neighboring microgrids," 50th International Universities Power Engineering Conference (UPEC), pp. 1–6, UK, 2015.
18. E. Pashajavid, F. Shahnia and A. Ghosh, "Interconnection of two neighboring autonomous microgrids based on small signal analysis," 9th International Conference on Power Electronics and ECCE Asia (ICPE-ECCE Asia), pp. 213–220, Korea, 2015.
19. E. Pashajavid, F. Shahnia and A. Ghosh, "A decentralized strategy to remedy the power deficiency in remote area microgrids," 50th International Universities Power Engineering Conference (UPEC), pp. 1–6, UK, 2015.
20. S. Bourbour, F. Shahnia and A. Ghosh, "Selection of a suitable microgrid to couple with an overloaded neighboring microgrid based on decision making," North American Power Symposium (NAPS), pp. 1–6, USA, 2015.
21. S.M. Chen and C.H. Wang, "A generalized model for prioritized multicriteria decision making systems," *Expert Systems with Applications*, vol. 36, Issue 3, pp. 4773–4783, 2009.
22. P.Ya. Ekel, J.S.C. Martini and R.M. Palhares, "Multicriteria analysis in decision making under information uncertainty," *Applied Mathematics and Computation*, vol. 200, Issue 2, pp. 501–516, 2008.
23. N. Fenton and W. Wang, "Risk and confidence analysis for fuzzy multicriteria decision making," *Knowledge-Based Systems*, vol. 19, Issue 6, pp. 430–437, 2006.
24. G. Campanella and R.A. Ribeiro, "A framework for dynamic multiple-criteria decision making," *Decision Support Systems*, vol. 52, Issue 1, pp. 52–60, 2011.
25. F. Shahnia, S. Bourbour and A. Ghosh, "Coupling neighboring microgrids for overload management based on dynamic multicriteria decision-making," *IEEE Trans. Smart Grid*, vol. 8, no. 2, pp. 969–983, 2017.

Distributed Agent-Based Coordinated Control for Microgrid Management

M.S. Rahman and A.M.T. Oo

Abstract Due to the ever-growing global concerns of climate changes and environmental issues, renewable energies along with the distributed energy resources (DERs) are becoming popular day by day for the sustainable operation of microgrids. Since a number of energy sources are connected to microgrids, multiple power electronic devices, such as inverters, are interfaced with those sources to support the DC–AC and AC–DC conversion. Multiple inverters installed with DERs are very effective in order to deliver adequate amount of energy from different sources to the end users. However, the control and management of multiple inverters adversely suffer from limited system stability due to the lack of proper coordination among different components of microgrids. Moreover, due to the requirements of sustainable operation of microgrids and intermittent characteristics of renewable energy, efficient coordinated control is still quite difficult to maintain the balance between the generations and demands. In order to deliver a reliable energy supply to microgrids, inverters need to be controlled in a more efficient way. Moreover, the integration of battery energy storage systems (BESSs) adds a new dimension to the reliable operation of microgrids which also possesses a difficulty to manage power sharing during the change in network configurations. In this chapter, a communication-assisted distributed multi-agent system (MAS)-based control scheme is proposed to effectively control and manage the inverter-dominated microgrids with solar photovoltaic (PV) systems and BESSs. The MAS establishes the coordination between the power sharing and energy management through agent communication and further ensures the sustainable operation of microgrids by effectively controlling the inverters.

Keywords Microgrid · Multi-agent system · Inverter · PV system · BESS · Distributed control

M.S. Rahman (✉) · A.M.T. Oo
Faculty of Science, Engineering and Built Environment,
School of Engineering, Deakin University, Geelong, VIC 3216, Australia

B. Azzopardi (ed.), *Sustainable Development in Energy Systems*,
DOI 10.1007/978-3-319-54808-1_4

1 Introduction

In recent years, among different renewable energy-based distributed generation (DG) units, the photovoltaic (PV) systems are increasingly being pursued as promising sources of energy due to their small relative size, noiseless operation, and feed-in-tariff [1, 2]. Moreover, the inverter-interfaced plug-in hybrid electric vehicles (PHEV)s equipping with battery energy storage systems (BESS)s have gained much attention as one of the popular distributed energy resources (DER)s due to the low environmental factors. These BESS units store extra amount of energy from the intermittent solar PV systems and thereby ensure a reliable energy supply to the end users when required. In this chapter, microgrids are considered as inverter-dominated networks which are subdivided into several subsystems depending on the number of DERs. Since several PV-based DG units and BESSs are installed in microgrids, multiple inverters are used for DC–AC conversion in order to effectively deliver electrical energy from different DERs to the residential and commercial consumers.

In microgrids, the renewable energy sources (RES)s and energy storage devices are connected to the feeders through power electronic inverters equipping with local controllers in which the coordination between various control equipment is very critical. Regardless the operating modes of microgrids, the inverters are managed and controlled using the local individual controllers within the subsystems. However, the control and management of multiple inverters suffer from a limited stability margin due to the lack of information exchange among different control equipment from different subsystems. In order to deliver and ensure reliable energy supply, the inverter-interfaced microgrids need to be controlled efficiently in a distributed fashion and this can be well achieved by the distributed multi-agent system (MAS) technology.

In the past few decades, a number of control techniques within the traditional frameworks have been used for the operation and management of microgrids [3–7]. Recently, a few multi-agent-based control techniques have been employed for the energy management of inverter-dominated microgrids [8–10]. In [11], a multi-agent-based coordinated control scheme along with a consensus algorithm is proposed for microgrid energy management within the energy internet framework. In [12], an intelligent agent-based control schemes is proposed for inverter-based BESSs management schemes of microgrids. In [13], an agent-based methodology for achieving coordination among the inverter-interfaced microgrids is proposed for improving the voltage profile. In [14], a multi-agent-based intelligent control scheme is designed for the operation and management of microgrids and in [15], a distributed multi-agent approach is proposed for microgrid control using feedback linearization techniques. In [16], a real-time agent-based intelligent control scheme is proposed to control and manage the power sources and loads in microgrids with BESSs. An autonomous software agent-based control scheme is proposed in [17] and [18] to manage energy resources in residential microgrids. Moreover, the introduction of communication systems in distributed controller design has brought the complexities and computational burdens [19]. A few more distributed multi-agent-based coordinated control for microgrid energy management are discussed in [20, 21].

However, the approaches, so far presented in this chapter, have some limitations of slow response which cannot handle fast varying network configurations and possessed limited stability margin with slow dynamics. Furthermore, most of the approaches brought numerous computational burdens which cause higher complexities to provide suitable and faster solutions to the given problems. In such cases, microgrids are difficult to coordinate with the present control schemes other than intelligent distributed control approach which has the ability of exchanging information among different subsystems. In order to overcome the drawbacks of the existing methods, a distributed agent-based scheme is designed to control and manage multiple inverters within microgrids. In this distributed multi-agent framework, agents take part in different activities for cost effective and reliable operation of residential microgrids. The multi-agent framework establishes the coordination among different subsystems and enables a communication network among various subsystems. The agents within the multi-agent framework exchange information with each other to decide their control actions in order to achieve the desired control objectives. In this chapter, a linear quadratic regulator (LQR) controller is designed for each inverter within the MAS framework for effectively controlling the inverters of microgrids. Since the agents are working in a distributed fashion, as a result, only the information from the neighboring subsystems is sufficient to coordinate the agent control actions. The performance of the proposed agent-based scheme is investigated on a microgrid system by considering various scenarios including faults and changes in atmospheric conditions. The results are also validated by comparing the proposed scheme to the conventional PI control scheme without MAS and communication.

2 Modeling of Microgrid System

Microgrids are localized group of various DERs including RESs and energy storage systems along with electrical loads. Microgrids can operate in both stand-alone mode (when disconnected from the utility grid) and grid connected mode (when connected to the utility grid). A schematic diagram of a typical microgrid model is shown in Fig. 1 which consists of total m-number of subsystems including renewable energy-based DGs (solar PV systems) and BESSs (PHEvs). The DGs comprise solar PV generation systems with associated power electronic inverters to the local bus using multiple parallel inverter topologies corresponding to the DG units and BESSs. In this microgrid model, pair of two nodes or buses is connected by each line, whereas each bus is connected to an inverter as well as rest of the system. The details of the physical device models of PV systems, BESSs, and loads for a typical microgrid are given in the following subsections.

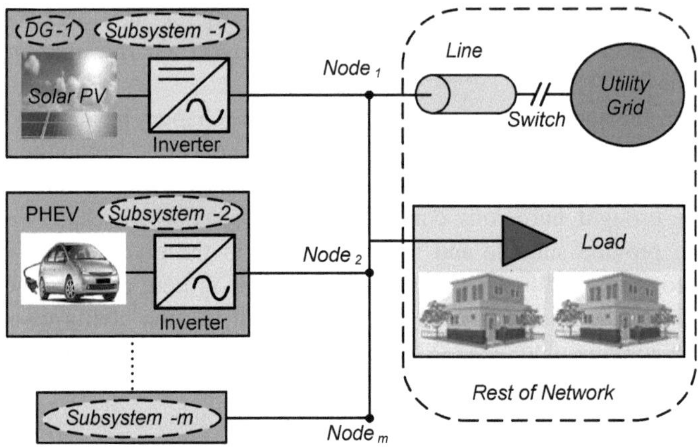

Fig. 1 Typical microgrid model

2.1 PV System Model

In solar PV systems, each PV module consists of a PV array, a single-phase inverter, a dc-link capacitor, and a filter inductor. A PV cell usually converts the solar irradiation directly from the sunlight into the electric energy. The schematic diagram of solar PV system connected to a microgrid is shown in Fig. 2 [1]. In this chapter, the dynamical model of a solar PV system in *dq*-frame is used, and the relevant equations are written as [1]:

$$\begin{aligned} \dot{I}_d &= -\frac{R}{L}I_d + \omega I_q - \frac{E_d}{L} + \frac{v_{pv}}{L}K_d \\ \dot{I}_q &= -\omega I_d - \frac{R}{L}I_q - \frac{E_q}{L} + \frac{v_{pv}}{L}K_q \\ \dot{v}_{pv} &= \frac{1}{C}i_{pv} - \frac{1}{C}I_d K_d - \frac{1}{C}IqK_q \end{aligned} \tag{1}$$

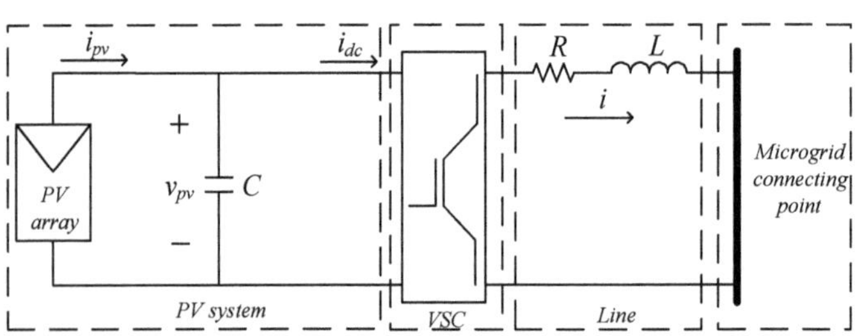

Fig. 2 Solar PV system connected to microgrid

where I_d and I_q are the dq-component of the inverter current, R is the line resistance, L is the inductance of the filter and line, ω is the angular frequency, E_d and E_q are the dq-axis voltage, C is the dc-link capacitance, v_{pv} is the voltage across the dc-link capacitor, i_{pv} is the output current of the PV array, K_d and K_q are the input switching signals.

2.2 Energy Storage Model

In this section, the mathematical model of the BESS is presented. A BESS usually consists of a battery, a voltage source converter (VSC), and an output filter. The most commonly used battery model is presented in [22] and its equivalent electrical circuit model connected to a microgrid system is shown in Fig. 3 [1]. The detail of battery elements such as resistors (R_0 and R_1), capacitor (C_1), and internal voltage (E_m) is found in [22]. The complete dynamical model of a BESS in dq-frame is written as [1]:

$$\begin{aligned}
\dot{I}_{sd} &= -\frac{R}{L}I_{sd} + \omega I_{sq} - \frac{E_d}{L} + \frac{v_{dc}}{L}M_d \\
\dot{I}_{sq} &= -\omega I_{sd} - \frac{R}{L}I_{sq} - \frac{E_q}{L} + \frac{v_{dc}}{L}M_q \\
\dot{I}_1 &= \frac{1}{T_1}\left(M_d I_{sd} + M_q I_{sq} - I_1\right)
\end{aligned} \tag{2}$$

where I_{sd} and I_{sq} are the dq-frame currents of battery connected inverters, I_l is the current through the battery, T_1 is the ratio of resistance and capacitance, M_d and M_q are the switching functions of generation modulation index in dq-frame, v_{dc} is the voltage across the capacitor.

In dq-frame, E_d is considered to be zero, and therefore, the real and reactive power outputs delivered from the solar PV systems and BESSs are written as:

$$\begin{aligned}
P_{\text{out}} &= E_q I_q \\
Q_{\text{out}} &= E_q I_d
\end{aligned} \tag{3}$$

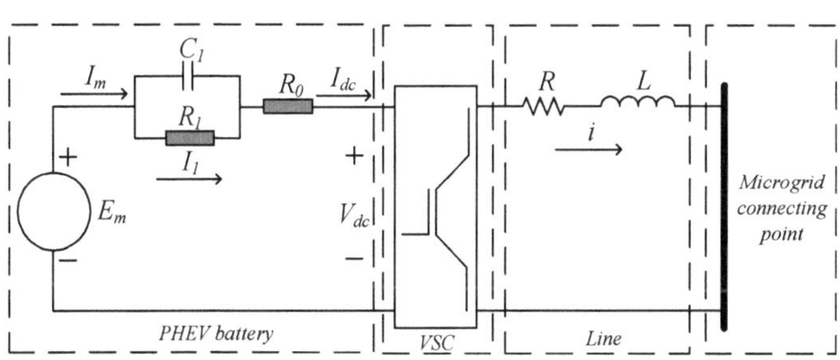

Fig. 3 BESS unit connected to microgrid

It is noted that the currents I_d and I_q are sufficient enough to regulate the real and reactive powers. The minimum state of charge (SOC) of the battery storage unit is considered as 30%.

2.3 Load Model

In this chapter, an aggregated composite ZIP (Z-constant impedance, I-constant current, P-constant power) load model is considered. The real and reactive power characteristics of relevant ZIP load model are written as [23, 24]:

$$\begin{aligned} P_L &= P_0\left(\frac{V}{V_0}\right)^{\alpha} \\ Q_L &= Q_0\left(\frac{V}{V_0}\right)^{\beta} \end{aligned} \tag{4}$$

where P_0 and Q_0 are the real and reactive power consumed at the reference voltage V_0, respectively. The exponential terms α and β depend on the type of the loads.

3 Agents and Multi-agent Systems

The background about the agents and multi-agent framework is discussed in the following subsections.

3.1 Agents

Agents are autonomous software components which provide an interoperable interface to an arbitrary system and behave similar to human agents working for some clients in pursuit of some specific agenda [25]. According to Wooldridge [26], an agent is merely 'a software entity that is present in an environment and is able to autonomously react to changes in that environment' which must be observable to, or alterable by, each intelligent agent. The environment may be a physical system, e.g., a power system, observable through sensors or PMUs, or a computing environment, e.g., data and computing resources obtained by messaging.

Each agent has the ability to interact with the neighboring agents over networked communications protocol which provides high-level autonomous software abstractions characterized by their behaviors and ontologies. Intelligent autonomous agents have two basic criteria [25]: knowledge base, i.e., each agent must be aware of the status of other agents and the whole environment; and message functionality, i.e., all the agents use their communication capability to exchange important message or information among themselves to achieve their common goal.

3.2 Multi-agent System (MAS)

A MAS framework is a distributed and coupled network composed of multiple, interacting agents within an environment [25]. Agents within a MAS work together to achieve a particular goal by solving a certain problem in a specific domain. The intelligence of an agent includes a few methodical, functional, procedural, and algorithmic searches, and a find and processing approach [27]. MASs are used in a wide range of applications require flexibility and adaptability for rapidly changing environments due to their distributed nature, modularity and ease of implementation [28]. The MAS offers an ideal means of achieving system integration by wrapping disparate systems as intelligent systems which often forms a distributed MAS. In a distributed MAS environment, agents collect local measurements or data from or send information to remote sites and receive or reproduce it at local sites. Recently, the MAS provides a more flexible way of increasing both the resilience of smart grids by combining top-down and bottom-up intelligent autonomous decision making in their communication and control architecture [29].

Agents and MAS technologies differ from existing systems and traditional engineering approaches since they have distributed characteristics which provides them dynamic adaptability and flexibility to cope with the physical systems. The potential advantages gained through this difference have motivated power engineers to explore applying MASs to power engineering problems [25]. The key component of the MAS is the communication principle where the intelligent agents can communicate with each other and each agent encapsulates a particular task or set of functionalities, in a similar way to object-oriented programming [30]. Agent use standard agent communication languages (ACLs) to cooperate and negotiate with neighboring ones to implement independent control actions.

4 MAS Framework for Sustainable Microgrids

In this chapter, a multi-agent system (MAS) framework is developed for a physical microgrid model. Two types of agents are designed within the MAS—(i) network agent (NA) and (ii) control agents (CA). The NA comprises the network information about the DERs including RESs and BESSs, and aggregated load models. Since the microgrid has several subsystems, a CA is designed for each inverter-interfaced DER within a local subsystem. The primary responsibility of the NA is to share the information to each CA of the local subsystems. Based on the information provided from the NA, the CA will adjust the control settings accordingly to control the inverters, and thereby, manage the power sharing for the sustainable operation of microgrids.

In this chapter, a directed graph (diagraph) theory is used to incorporate the bidirectional communication and coordination of agents. The communication-assisted agents obviates the requirements for a central controller and complex communication network to improve the system reliability. The communication diagraph for the agent framework can be presented by the following graph theory:

$$G = (V, A_G) \tag{5}$$

where V is the set of buses or nodes connected to the inverter-interfaced DERs within individual subsystems and A_G is the relevant adjacency matrix $A_G = [a_{ij}] \in R^{N \times N}$. This graph provides the available bidirectional communication links among different agents through which they can coordinate their activities. For each subsystem, it is desirable that the power injection from the inverters can be shared properly to their nominal values to maintain the frequency and voltage, respectively. The NA can communicate with the CAs through the diagraph and each local CA can further exchange information with its neighboring CA via the same diagraph. The bidirectional communication topology is shown in Fig. 4 in which the dashed line represents the communication link among various agents. In this chapter, it is considered that the communication channel is ideal in nature, i.e., there is no time delay in agent communication. The communication network that superimposed on the physical layer of the microgrid is assumed to be strongly connected without any data loss [31, 32].

In the proposed MAS framework, the agents continuously perform their internal activities based on the real-time network information and share the information to others. Moreover, each CA has the full access to the local information of neighboring CAs to coordinate the control activities and, thereby, respond to the messages already being sent. As a result, the agents can enjoy the benefits of using local as well neighboring information which make the system fully distributed.

Fig. 4 Communication topology among agents

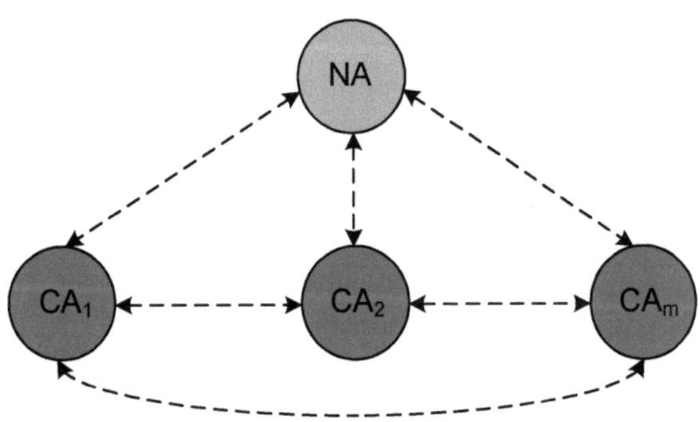

5 Agent-Based Controller Design

The Pulse-width modulation (PWM) technique is used to generate the switching pulses to drive the IGBT switches for controlling the inverter in such a way that the output signal is sinusoidal. The main control objective of the CA is to obtain the fast transient response with a minimum error. In an agent-based inverter control scheme, the NA provides the information of load voltage and inverter current obtained from the voltage and current sensors to the CA. After receiving the information from the NA, the CA will track the reference output and hence, control the *dq*-component of the inverter current. The close-loop system within the CA is written as:

$$u = f(F, F^{*}, y, y^{*}, u(\varepsilon)) \tag{6}$$

where F represents all structural online information, F^{*} is the estimation of F, y is the measured output, $y*$ is the reference output trajectory, ε is the tracking error signal, and u is the control input to the system. A LQR control scheme is designed within each CA for the control action and the controller design algorithm is discussed in the following subsection.

5.1 LQR Controller Design

The LQR control scheme is designed to determine the optimal control law for minimizing the cost function. In order to design the LQR controller for both PV systems and BESSs model, the linearized form of Eqs. (1) and (2) is written as:

$$\begin{aligned} \dot{x}(t) - Ax(t) + Bu(t) \\ y(t) = Cx(t) \end{aligned} \tag{7}$$

where $x \in R^{n}$ is the state vector of matrix A, $u \in R^{m}$ is the input vector of matrix B, and y is the output vector of matrix C. Therefore, after linearization, Eqs. (1) and (2) are written as:

- for PV system:

$$\dot{x} = \begin{bmatrix} \dot{I}_d \\ \dot{I}_q \\ \dot{v}_{pv} \end{bmatrix} = \begin{bmatrix} -\frac{R}{L} & \omega & 0 \\ -\omega & -\frac{R}{L} & \omega \\ 0 & 0 & \frac{1}{C} \end{bmatrix} \begin{bmatrix} I_d \\ I_q \\ v_{pv} \end{bmatrix} + \begin{bmatrix} \frac{v_{pv}}{L} & 0 \\ 0 & \frac{v_{pv}}{L} \\ -\frac{I_d}{C} & -\frac{I_q}{C} \end{bmatrix} \begin{bmatrix} K_d \\ K_q \end{bmatrix}$$

- for BESS system:

$$\dot{x} = \begin{bmatrix} \dot{I}_1 \\ \dot{I}_d \\ \dot{I}_q \end{bmatrix} = \begin{bmatrix} -\frac{1}{T_1} & 0 & 0 \\ 0 & -\frac{R}{L} & \omega \\ 0 & -\omega & -\frac{R}{L} \end{bmatrix} \begin{bmatrix} I_1 \\ I_d \\ I_q \end{bmatrix} + \begin{bmatrix} \frac{I_d}{T_1} & \frac{I_q}{T_1} \\ \frac{v_{dc}}{L} & 0 \\ 0 & \frac{v_{dc}}{L} \end{bmatrix} \begin{bmatrix} M_d \\ M_q \end{bmatrix}$$

In all cases, the output of the controller can be written as:

$$y = [I_d \quad I_q]$$

The performance index of the LQR controller is chosen as:

$$\min_{\text{Klqr}} J = \int_0^\infty \left(x^{\mathrm{T}}(t)Qx(t) + u^{\mathrm{T}}(t)Ru(t)\right)\mathrm{d}t \tag{8}$$

where Q is a semi- or positive definite state-weighting matrix normally chosen as $Q^{\mathrm{T}} \geq 0$ and R is a real symmetric positive definite control weighting matrix usually chosen as $R^{\mathrm{T}} > 0$. Since the key focus of the LQR controller design is to find the optimal control law with minimum J, the optimal control law can be written as:

$$u(t) = K_{\text{lqr}}x(t) \tag{9}$$

For reduced cost function, the optimal gain matrix of the LQR controller can be written as:

$$K_{\text{lqr}} = R^{-1}B^{\mathrm{T}}P \tag{10}$$

where P is a positive definite unique solution of the following control algebraic Riccati equation (ARE):

$$A^{\mathrm{T}}P + PA - PBR^{-1}B^{\mathrm{T}}P = 0 \tag{11}$$

The designed controller within the CA is used to control the parameters I_d and I_q to make them rapidly track their respective current commands I_{dref} and I_{qref} which syntheses the modulating functions according to the real power reference signal P_{ref} and reactive power reference signal Q_{ref}, respectively. Since the NA shares a continual streaming of measured voltage and current information, the CAs can adjust the gain of the LQR controllers according to the changes in system configurations. The CA uses the control signal to drive the PWM generator by controlling the inverter. A proper gain selection of the controller will ensure accurate tracking and short transients in the output parameters.

5.2 *Selection of Weighting Matrices*

In all LQR design algorithm, a key issue is to choose the state and control weighting matrices in such a way that the closed-loop system performances can satisfy the desired control objectives. In order to avoid the computational burdens and complexities of tradition LQR design, the weighting matrices are chosen in the following way to obtain the satisfactory results:

$$\begin{aligned} Q &= C^{\mathrm{T}}C \\ R &= 1 \end{aligned} \tag{12}$$

The designed controller within the CA is used to maintain the balance between the generation and demand, thereby, manage the effective power sharing for sustainable microgrid operation. The performance of the proposed scheme is evaluated on a microgrid and illustrative examples with results are discussed in the following section.

6 Performance Evaluation and Case Studies

In order to evaluate the performance of the proposed scheme, a typical microgrid system, very similar to a practical system, as shown in Fig. 5 is considered in this chapter. This microgrid system operates at 50 Hz and consists of two inverter-interfaced solar PV units, one inverter-interfaced BESS unit, and aggregated electrical load. This system is considered to be running in stand-alone mode and since the solar PV generation is intermittent in nature, it is necessary to maintain the balance between the generation and the demand. In such case, BESS units are utilized to provide supply to the loads and the inverters need to be controlled in a proper way. In this microgrid system, two PV units, each of 8 kW,

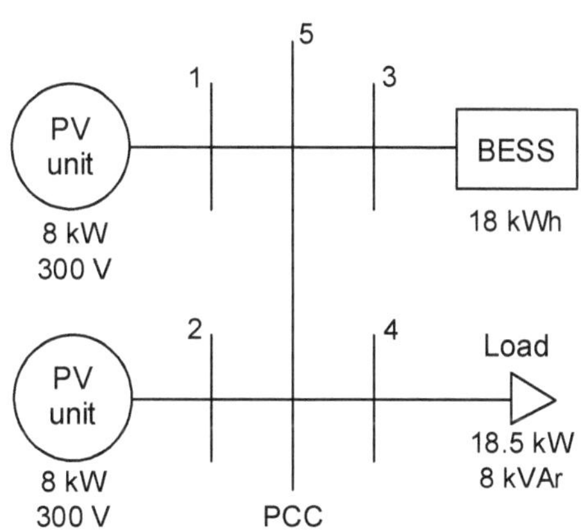

Fig. 5 Single-line diagram of microgrid system

300 V, are connected to bus-1 and bus-2 to supply to the aggregated electrical load of 18.5 kW, 8 kVAr, at bus-4, and the excess power is stored in the BESS unit (18 kWh) connected at bus-3. In this system, bus-5 is the PCC since all the components are commonly connected to it. The dc-link capacitor value of this system is 200 μF, and the line resistance is 0.2 Ω with the inductance of 12 mH. The switching frequency of each inverter is considered as 4 kHz. The value of the solar irradiation is considered as 1000 Wm^{-2} and the temperature as 25 °C.

When there is a change in the system configuration, perhaps due to faults or weather conditions, the system behavior changes significantly. As a result, some fluctuations appear in the voltage and current profiles which can be minimized by regulating the inverter switches through proper control actions. In order to demonstrate the superiority of the proposed agent-based scheme, the following two case studies are considered for the performance evaluation, and finally, the proposed scheme is compared to a conventional PI control scheme without MAS. The brief description of a typical PI controller can be seen in [33].

- Case-1: Performance evaluation under faulted condition
- Case-2: Performance evaluation for change in atmospheric conditions.

6.1 Performance Evaluation Under Fault Condition

In this case study, a short-circuit fault is applied at bus-1 at 0.2 s, and it is cleared at 0.3 s. In such condition, the PV unit-1 will not supply any power into the load and the BESS unit supplies energy. In such condition, it is essential to regulate the inverter switches to control the current in order to maintain stable load voltage and current which is done perfectly by the agent coordination. The NA sends the information of the measured signals to the CA which will take the main control decisions to control the inverters.

The load voltage and current profile for this case study are shown in Figs. 6 and 7, respectively. From Figs. 6 and 7, it is seen that the load voltage and current have some unstable fluctuations immediately after clearing the fault when conventional control scheme is used. On the other hand, from Figs. 6 and 7, it is seen that the proposed agent-based scheme maintains a post-fault steady-state voltage and current as soon as the fault is cleared. Moreover, the load voltage and current waveforms are nearly smooth due to the proper control actions taken by the CA, whereas they are producing some ripples when conventional PI control scheme is used. That means, the designed MAS scheme maintains the stable load voltage and current profiles in a better way as compared to the conventional PI control scheme without MAS and communication.

If the clearing of the fault is not performed at the zero crossing of the voltage, the voltage profile fluctuates more when the fault is cleared, and it is shown in Fig. 8. From Fig. 8 it is seen that, the proposed scheme performs well as compared to the conventional scheme. The real and reactive powers shared from the BESS unit are

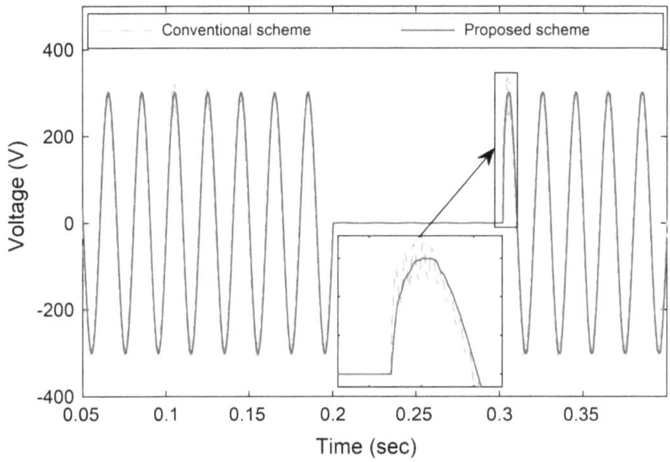

Fig. 6 Load voltage profile for short-circuit fault at 0.2 s

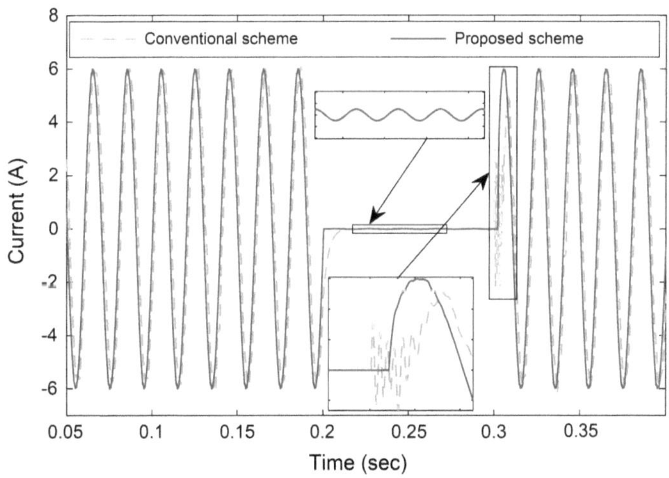

Fig. 7 Load current for short-circuit fault at 0.2 s

shown in Figs. 9 and 10, respectively, in which it is seen that the real and reactive powers are settled down very quickly at the post-fault period when the proposed scheme is used. On the other hand, it is also seen that the real and reactive powers oscillates and take longer time to settle at the post-fault period when the conventional scheme is used. That means, the proposed agent-based scheme brings the system to a steady-state condition very quickly after the fault is cleared and this ensures the effectiveness of the proper utilization of the BESS units through the agents.

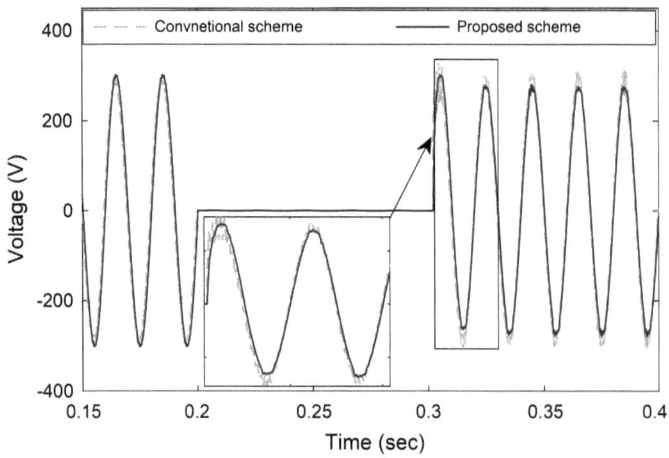

Fig. 8 Voltage profile for fault cleared other than zero crossing of voltage

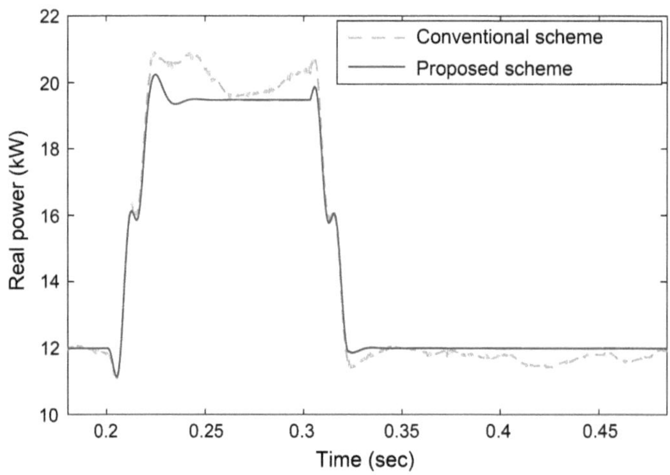

Fig. 9 Real power sharing during faulted condition

6.2 Performance Evaluation for Changes in Atmospheric Conditions

In this case study, the simulations are carried out for a change in the atmospheric conditions. In practice, since the direct sunlight is not available all the time in a day, the solar irradiation and temperature are variable. As a result, the output voltage and power of the PV array change significantly. In order to observe the impacts of change in the atmospheric condition to the microgrid, the solar irradiation is

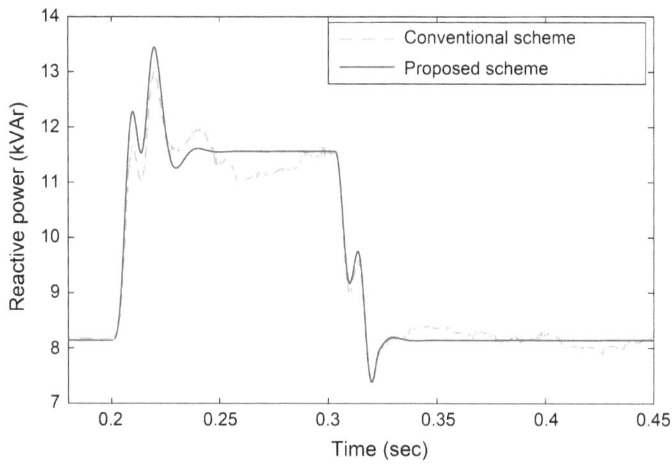

Fig. 10 Reactive power sharing during faulted condition

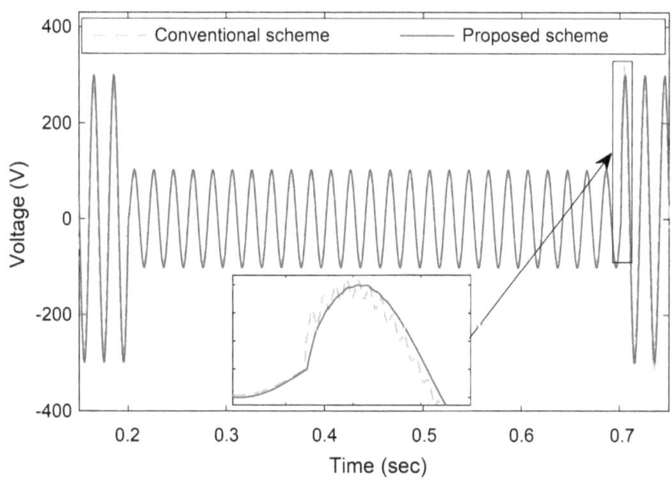

Fig. 11 Load voltage profile for changes in solar irradiation

changed from 1000 to 550 Wm^{-2} at 0.2 s and continues till 0.7 s with a total duration of 0.5 s. It is considered that, after 0.7 s, the system is again running normal at standard atmospheric conditions with solar irradiation 1000 Wm^{-2}.

The load voltage profile for this change in atmospheric condition is shown in Fig. 11 in which it can be seen that the voltage profile varies significantly and contain much ripples when the conventional PI control scheme is used, whereas the voltage appears nearly sinusoidal when the proposed MAS scheme is used. The corresponding real and reactive power outputs are shown in Figs. 12 and 13, respectively. From Figs. 12 and 13, it can be seen that their amplitudes change

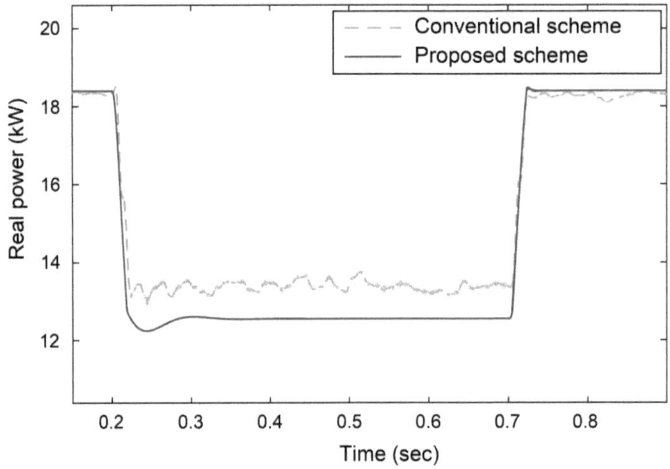

Fig. 12 Real power output for changes in solar irradiation

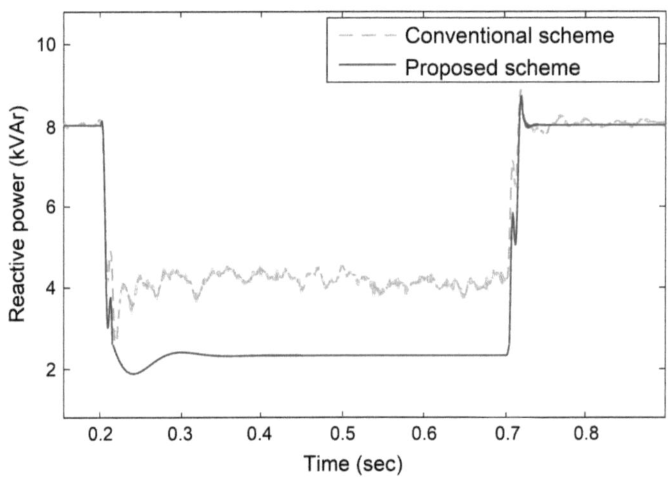

Fig. 13 Reactive power output for changes in solar irradiation

when the solar radiation is drops down from 1000 to 550 Wm^{-2} at 0.2 s. It is also seen that, the real and reactive powers return to the previous condition at 0.7 s and become much improved when proposed agent-based scheme is used as compared to the conventional control scheme. From the simulation results, it is obvious that the proposed scheme provides a good steady-state performance under the change in the weather conditions.

7 Conclusion

In this chapter, an innovative multi-agent framework is developed for the sustainable operation of microgrids. The agents dynamically adapt to the changes in the system configuration and thereby, take appropriate actions for enhancing the stability and reliability of the microgrid. From the simulation results, it is clear that the proposed MAS scheme performs much better than the traditional control scheme and all operations are performed in a smarter way based on the information exchange among various agents over the network.

References

1. M. Mahmud, H. Pota, and M. Hossain, "Nonlinear current control scheme for a single-phase grid-connected photovoltaic system," IEEE Trans. on Sustainable Energy 5(1):218–227, 2014.
2. A. Zahedi, "Development of an economical model to determine an appropriate feed-in tariff for grid-connected solar PV electricity in all states of Australia," Renewable and Sustainable Energy Reviews 13(4):871–878, 2009.
3. J. Yang, Z. Zeng, Y. Tang, J. Yan, H. He, and Y. Wu, "Load frequency control in isolated micro-grids with electrical vehicles based on multivariable generalized predictive theory," Energies 8(3):21–45, 2015.
4. J. Schiffer, R. Ortega, A. Astolfi, J. Raisch, and T. Sezi, "Conditions for stability of droop-controlled inverter-based microgrids," Automatica 50(10):2457–2469, 2014.
5. Y. Mou, H. Xing, Z. Lin, and M. Fu, "Decentralized optimal demand-side management for PHEV charging in a smart grid," IEEE Trans. on Smart Grid 6(2):726–736, 2015.
6. A. Arancibia, K. Strunz, and F. Mancilla-David, "A supervisory energy management control strategy in a battery/ultracapacitor hybrid energy storage system," IEEE Trans. on Transportation Electrification 1(3):223–231, 2015.
7. T. Caldognetto, P. Tenti, A. Costabeber, and P. Mattavelli, "Improving microgrid performance by cooperative control of distributed energy sources," IEEE Trans. on Industry Applications 50(6):3921–3930, 2014.
8. W. Liu,W. Gu, W. Sheng, X. Meng, Z.Wu, and W. Chen, "Decentralized multi-agent system-based cooperative frequency control for autonomous microgrids with communication constraints," IEEE Trans. on Sustainable Energy 5(2):446–456, 2014.
9. C. Colson and M. Nehrir, "Comprehensive real-time microgrid power management and control with distributed agents," IEEE Trans. on Smart Grid 4(1):617–627, 2013.
10. T. Tsuji, T. Oyama, T. Hashiguchi, T. Goda, K. Horiuchi, S. Tange, T. Shinji, and S. Tsujita, Autonomous Decentralized Voltage Profile Control Method in Future Distribution Network using Distributed Generators. InTech, 2011, ch. 11:193–220.
11. Q. Sun, R. Han, H. Zhang, J. Zhou, and J. Guerrero, "A multiagent-based consensus algorithm for distributed coordinated control of distributed generators in the energy internet," IEEE Trans. on Smart Grid 6(6):3006–3019, 2015.
12. C.-H. Yoo, I.-Y. Chung, H.-J. Lee, and S.-S. Hong, "Intelligent control of battery energy storage for multi-agent based microgrid energy management," Energies 6(10):4956–4979, 2013.
13. E. Polymeneas and M. Benosman, "Multi-agent coordination of DG inverters for improving the voltage profile of the distribution grid," in IEEE PES General Meeting 2014:1–5.

14. M. Kouluri and R. Pandey, "Intelligent agent based micro grid control," in International Conference on Intelligent Agent and Multi-Agent Systems (IAMA) 2011:62–66.
15. A. Bidram, A. Davoudi, F. Lewis, and J. Guerrero, "Distributed cooperative secondary control of microgrids using feedback linearization," IEEE Trans. on Power Systems 28 (3):3462–3470, 2013.
16. M.-T. Kuo and S.-D. Lu, "Design and implementation of real-time intelligent control and structure based on multi-agent systems in microgrids," Energies 6(11): 6045–6059, 2013.
17. A. Werth, N. Kitamura, and K. Tanaka, "Conceptual study for open energy systems: Distributed energy network using interconnected DC nanogrids," IEEE Trans. on Smart Grid 6(4):1621–1630, 2015.
18. M. D. Galus, R. A. Waraich, and G. Andersson, "Predictive, distributed, hierarchical charging control of PHEVs in the distribution system of a large urban area incorporating a multi agent transportation simulation," in Power Systems Computations Conference (PSCC) 2011:1–7.
19. L. Herrera, E. Inoa, F. Guo, J. Wang, and H. Tang, "Small-signal modeling and networked control of a PHEV charging facility," IEEE Trans. on Industry Applications 50(2):1121–1130, 2014.
20. J. Oyarzabal, J. Jimeno, J. Ruela, A. Engler, and C. Hardt, "Agent based micro grid management system," in International Conference on Future Power Systems 2005:1–6.
21. Z. Xiaoyan, L. Tianqi, and L. Xueping, "Multi-agent based microgrid coordinated control," Energy Procedia 14:154–159, 2012.
22. M. Ceraolo, "New dynamical models of lead-acid batteries," IEEE Trans. on Power Systems 15(4)1184–1190, 2000.
23. M. S. Rahman, M. A. Mahmud, H. R. Pota, and M. J. Hossain, "Distributed multi-agent scheme for reactive power management with renewable energy," Energy Conversion and Management 88:573–581, 2014.
24. T. Aziz, T. K. Saha, and N. Mithulanathan, "Distributed generators placement for loadability enhancement based on reactive power margin," 9th International Power & Energy Conference (IPEC) 2010: 740–745.
25. M. S. Rahman, "Distributed Multi-Agent Approach for Enhancing Stability and Security of Emerging Smart Grids," PhD Thesis, The University of new South Wales, 2014.
26. W. M. and W. G., Intelligent agents. MIT press, Cambridge, 1999, ch. Multiagent systems, pp. 3–51.
27. S. D. J. McArthur, E. Davidson, J. Hossack, and J. McDonald, "Automating power system fault diagnosis through multi-agent system technology," in 37th Annual Hawaii International Conference on System Sciences 2004: 1–8.
28. F. LN, "Entertaining agents: a sociological case study" in 1st international conference on autonomous agents, 1997: 122–129.
29. M. Rahman, M. Mahmud, H. Pota, M. Hossain, and T. Orchi, "Distributed Multi-Agent-Based Protection Scheme for Transient Stability Enhancement in Power Systems," Int. Journal of Emerging Electric Power Systems 16 (2):117–129, 2015.
30. S. McArthur, E. Davidson, V. Catterson, A. Dimeas, N. Hatziargyriou, F. Ponci, and T. Funabashi, "Multi-agent systems for power engineering applications part I: Concepts, approaches, and technical challenges," IEEE Trans. on Power System 22:1743–1752, 2007.
31. M. S. Rahman, M. A. Mahmud, A. M. T. Oo, and T. F. Orchi, "Distributed agent-based control scheme for single-phase parallel inverters in microgrids with photovoltaic systems," Australasian Universities Power Engineering Conference (AUPEC) 2015: 1–6.
32. M. Rahman, M. Mahmud, H. Pota, and M. Hossain, "A multi-agent approach for enhancing transient stability of smart grids," Int. Journal of Electrical Power & Energy Systems 67:488–500, 2015.
33. D. Zammit, C. Spiteri Staines, M. Apap, "Comparison between PI and PR Current Controllers in Grid Connected PV Inverters", Int. Journal of Electrical, Computer, Energetic, Electronic and Communication Engineering 8(2):221–226, 2014.

Sustainable Development of Energy Systems in the Baltic Region

Jurijs Spiridonovs and Olga Bogdanova

Abstract This Chapter addresses the challenges of the energy sector in the Baltic States, which are quite similar by size and the path of economic development; however unique in their energy policies at the same time. The Chapter provides wide information on the situation in power generation, infrastructure for power and gas, explains the situation on formation the energy markets in the region, as well as touches the issues of energy security, which are particularly sensitive for the Baltic States being still insufficiently connected to the rest of the European Union. Particular attention is paid to sustainability issues and increase of share of renewable energy sources in energy mix, reflecting the experience of the Baltic States in applying the state support mechanisms. The Chapter provides a common picture of the recent developments in the energy sector of the region and demonstrates the most important projects to be implemented in the coming years, drawing future perspectives and conclusions. The best practices and the experience gained from the implemented policies of the Baltic States could be also applied in other regions.

Keywords The Baltic States · Energy policy · Renewable energy sources · Energy infrastructure · Energy markets · Energy security

J. Spiridonovs (✉)
Economics and Finance Department, RISEBA University of Business, Arts and Technology, Riga, Latvia
e-mail: j.spiridonovs@gmail.com

O. Bogdanova
International Business and Customs Institute, Riga Technical University, Riga, Latvia
e-mail: nameolga@yahoo.co.uk

B. Azzopardi (ed.), *Sustainable Development in Energy Systems*,
DOI 10.1007/978-3-319-54808-1_5

1 Introduction

The Baltic States (Lithuania, Latvia and Estonia) are unique in many ways: still hugely dependent on one-source gas supplies, operating in non-EU electricity frequency system, having comparatively low level of interconnections between the other countries in the region and the largest functioning underground gas storage in the region. The issue of energy security still determines the development path in the region, making impact on development scenario in the mentioned economies. Despite many similarities, these three countries do represent versatile aspects of energy policy: nuclear energy developments in Lithuania, lignite as the main energy source in Estonia and more than 38% of energy coming from renewable energy sources in Latvia.

One of the main aims of the EU internal market legal regulation, so-called third energy market package (electricity and gas), is to develop well-functioning energy market, ensuring integrated net of infrastructure within entire EU territory. The idea foresees to create capable enough interconnections between all the EU member states therewith eliminating considerable dependence of particular countries on energy imports from non-EU countries, as well as ensure that the interconnected EU market operates according to harmonized EU principles and regulations. A closer integration of networks is particularly important to the states considered to be "energy islands", when electricity and gas networks are not connected with the other EU member states.

Necessity for government to execute certain economic policy is determined by the new challenges appearing from changes in the economic processes, as well as the need to establish the appropriate mechanisms for addressing these changes, identifying the necessary resources for implementing selected governmental economic policy. In case of the Baltic Countries, this also means, on the one hand, overcoming of historic heritage and, on the other hand, reacting to the modern economic and political questions: dramatic decrease of population, slow economic growth rates and risks of political insecurity in the wider region.

Moreover, transition process to market-based relations in the field of energy has its impact on all the aspects of the energy policies in the Baltic States, changing the traditional perception of industries and households of the services of energy companies.

The goal of this *chapter* is to explore basic facts and challenges of the Baltic region for its sustainable development in the perspective of EU Energy Union project.

2 The Baltic States: Basic Facts

Estonia, Latvia and Lithuania are located in the Eastern Europe and are rather different from the rest of Europe in terms of energy policy due to their current situation in energy sector and the historical heritage. All the three countries (Baltic region) of the European Union have population of 6.2 million people, territory of 175,000 km^2, GDP of about 15 thousand USD dollars per capita and the annual energy consumption of 748,000 PJ (Fig. 1).

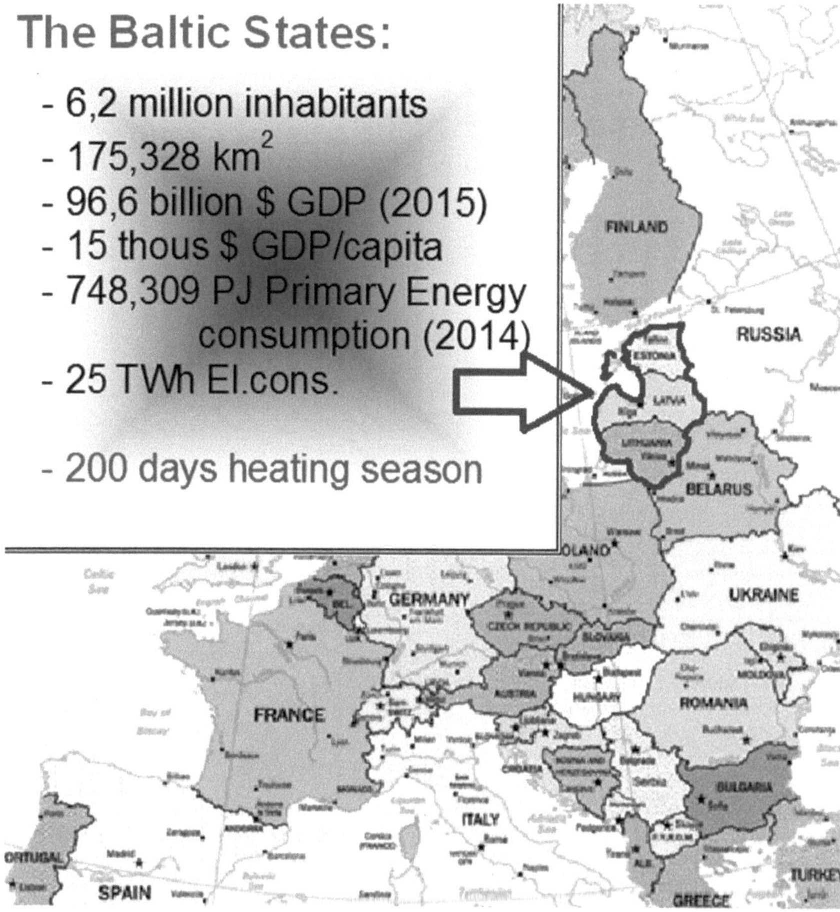

Fig. 1 Basic facts on the Baltic States. *Source* Eurostat

Practically, it means that energy market in terms of territory, number of final consumers and intensity of energy consumption is relatively small. Another peculiarity of the region is having about 200 days of heating season due to climatic conditions, making it reasonable to produce above 35% of electricity in CHP mode, as electric power generation in condensation regime is less cost-efficient.

Level of economic development of the Baltic State is similar also in the aspect of public spending in comparison with GDP (see Fig. 2): these countries are rather dependent on the support from the EU structural funds and centralized financial instruments. Therefore, in pursuit of development of the necessary energy infrastructure, the countries heavily rely on availability of EU financing, e.g. *Connecting Europe Facility* (CEF), and similar ways of funding. Within the first CEF financing call in October 2014, the Baltic region has received grants for implementation of energy infrastructure projects for 506,49 million EUR, which is 78% of all the CEF allocated budget during the first call.

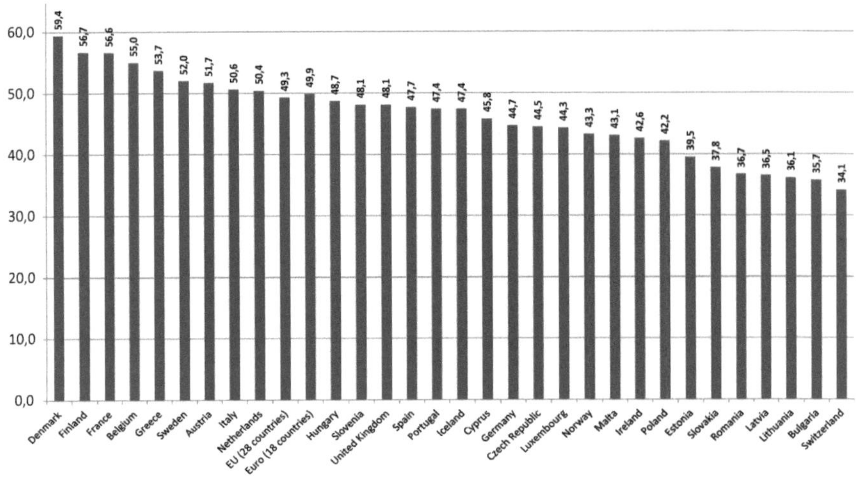

Fig. 2 Budget expenditure in EEZ counties compared to GDP, 2014. *Source* Eurostat

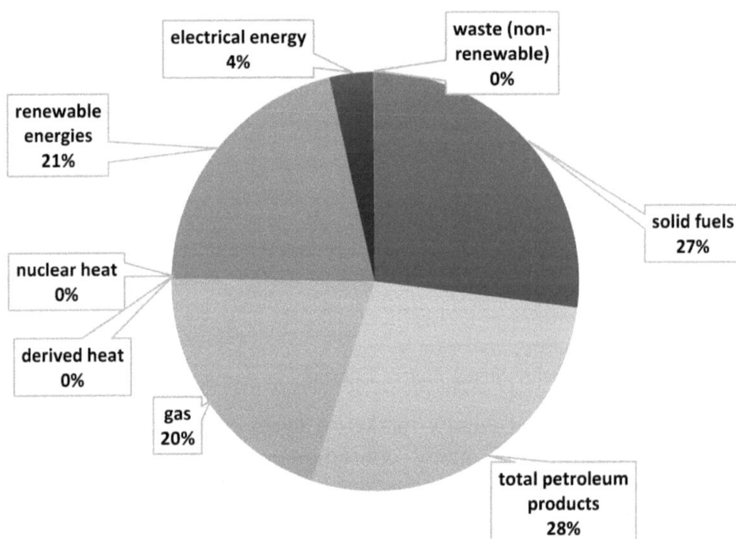

Fig. 3 Gross inland energy consumption in the Baltic States in 2014. *Source* Eurostat

Generally, the primary energy mix of the Baltic States is relatively good diversified. As it is demonstrated in Fig. 3, petroleum products, solid fuels, renewable energies and gas take the leading positions of the Baltic energy consumption basket. The three Baltic States have high share of renewable energy sources in the energy mix in comparison with other countries of the EU. Furthermore, the region has quite ambitious goals for further RES power generation development—Latvia, which is the second greenest country among all the EU member states, has RES target 40%, Estonia—25% and Lithuania—23%.

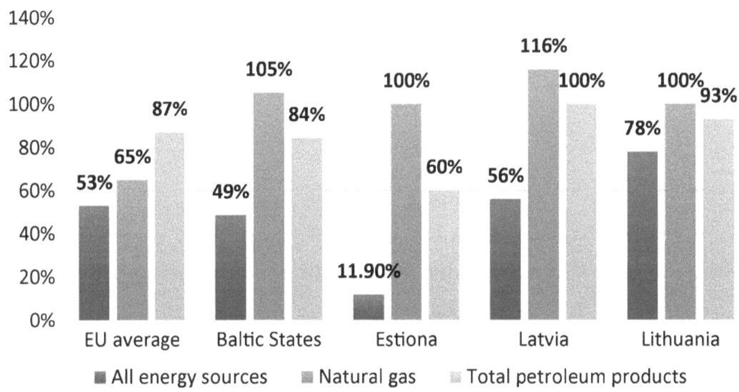

Fig. 4 Energy sources import dependency, 2013. *Source* European Commission

Within the last years, the decentralized power generation from RES in the Baltic States has been rapidly increasing, reaching a growth rate of more than 11% compared to 1995. In times when many EU countries find it challengeable meeting 20% RES target in 2020, the Baltic States demonstrated excellent results already in 2014 and all of them are to fulfil the RES share obligation by 2020.

At the same time, the Baltic region still remains being highly dependent on energy imports, which make the region vulnerable to external economic shocks and sensitive to geopolitical risks (Fig. 4).

In 2013, the general energy source import dependency in Lithuania and Latvia was higher than the EU average, but for oil and petroleum products (please notice that Lithuania actually imports oil and produces petroleum)—considerably higher than the EU average. The critical situation was in gas sector, having all the natural gas imported to all the three Baltic States. Moreover, the gas supplies ensured from the non-EU country Russia were dominated.

Therewith, the biggest challenges of energy policy are related to increasing energy supply security in electricity and gas sectors.

3 Power Generation

Currently, the Baltic region in general is self-sufficient in terms of installed power generation capacities; however, situation differs from country to country. Due to the market conditions and the available generation capacities, Estonia has surplus of generated power (generating 128.4% of domestic consumption), when Latvia and even more Lithuania are considerably import-dependent (Table 1).

Nevertheless, generally the energy mix in electricity power generation is diverse both by technology and by energy source. Efficient cooperation between transmission system operators makes the power system operation stable and reliable.

In 2010, the situation in power generation of the Baltic States substantially changed due to closing of the Lithuanian Ignalina Nuclear Power Plant having two

Table 1 Self-sufficiency in primary energy sources and electricity generation

Self-sufficiency	Primary energy sources		Electricity sector	
year	2013 (%)	2014 (%)	2013 (%)	2014 (%)
Latvia	49	55	82.10	68.90
Lithuania	24	25	40.70	36.60
Estonia	94	97	137.00	128.40
The Baltic States	56	59	87	78

Source Eurostat

units of 1500 MW of installed capacity. The Ignalina NPP ensured about 40% ($\sim$10 TWh) of all the generated power in the region, so its decommissioning turned the region from export-oriented to import-oriented.

The power system of Estonia has at present 2.7 GW of generation capacity installed and that capacity is sufficient to cover peak loads. In 2013, power in Estonia was mainly produced by large thermal power plants (Eesti, Balti and Iru), with a total generation of 11,892 GWh/year and renewable sources, mainly wind power, accounting for 451 GWh/year. Regarding the generation adequacy scenario developed by ENTSO-E, generation adequacy of the Estonian system should not be at risk up to 2020. The most important investments from the security of supply point of view were construction of a second electricity interconnection between Finland and Estonia with a capacity of 0.65 GW and construction of the new power plant in Estonia with 0.25 GW as an emergency reserve. Nevertheless, the Estonian network may occasionally experience lower voltages compared with the other two Baltic State power systems (Latvia and Lithuania), especially in cases, when the interconnections between Estonia and Finland (named Estlinks) are under heavy load conditions. Reactive compensations may be needed, particularly around the area of Tallinn. Even though the existing installed wind turbine capacity of Estonia is the highest among the three Baltic States, this is still not enough to give Estonia the same strategic market position as Latvia has.

The power system of Latvia has advantages in comparison with its neighbour countries due to the high ratio of renewable energy—mainly hydro—in its electricity generation mix. The most of electric power in Latvia is generated by CHPs and HHPs. The two CHPs in Riga (Rigas CHP1 and Rigas CHP2) ensure the base load and have the installed capacity of 1.025 GW. The cascade of hydropower plants (Rigas HPP, Kegums 1 & 2 HPP and Plavinas HPP) on the Daugava River is used as a reserve for covering demand during peak hours and has capacity of 1.575 GW. The only limitation for the power generation by the HPPs is water inflow in Daugava. There are a number of small gas CHPs in Latvia having capacity of 130 MW. The highest increase in generation capacities is expected in wind and biomass/biogas fields. Currently, Latvia has around 80 MW of wind capacity installed, which according to conservative scenario is going to increase to around 500 MW in 2025. The biomass/biogas generation type dominates for the coming years, and Latvian TSO is expecting that biomass/biogas generation will reach 130–

190 MW. The generation of solar is very insignificant. In 2013, the Latvia's HPPs produced 2912 GWh/year, CHPs—2869 GWh/year, but wood and wind power plants generated 119 GWh/year.

After closing the Ignalina NPP, the generation capacity of Lithuania is explicitly negative and Lithuania heavily relies on energy import. The Lithuanian generation capacity is 4323 MW, consisting of hydropower and pumped storage power plants for 1066 GWh/year (Kaunas and Kruonis plants); gas, black fuel or oil for 2615 GWh/year (Vilnius, Mazeikiu, Kaunas and Elektrenai power plants); and wind produced 649 GWh/year. A new nuclear power plant in Visaginas with capacity of 1350 MW is still under consideration. However, a range of important issues (such as safety and cost efficiency) are to be resolved before the project could be considered for being realized.

Despite the power generation deficit, the internal electricity network of Lithuania is quite powerful and can sustain large quantity of imports. The hydropump station is important for the shifting of generation resources and thus plays a key role in reducing the system marginal cost. Moreover, from 2016 Lithuania has new interconnections with Sweden and Poland (700 and 500 MW, respectively), ensuring possibility to import the power deficit in a cost-efficient way.

4 Gas Market Developments

Natural gas takes 20% of the primary energy mix of the Baltic States. The total gas consumption of the region was 2739.2 mcm in 2014, including 1294 mcm of gas consumption in Latvia, 912.2 mcm in Lithuania and 533 mcm in Estonia. More than 80% of gas is used for power and heat generation, households use about 12% of gas, but gas consumption in industrial sector is relatively small (however, almost half of gas consumption in Lithuania goes to one fertilizer manufacturer).

Generally, the gas consumption within the last decade has been constantly decreasing. The volume of consumed gas is strongly correlated with the following factors:

- Outside temperature and climate conditions;
- Overall development of economy;
- Shift from fossil to renewable energy sources for power and heat generation;
- Energy efficiency measures (Directive 2012/27/EU of the European Parliament and of the Council of 25 October 2012 on energy efficiency, amending Directives 2009/125/EC and 2010/30/EU and repealing Directives 2004/8/EC and 2006/32/EC),[1] etc.

[1]The 2012 Energy Efficiency Directive establishes a set of binding measures to help the EU reach its 20% energy efficiency target by 2020. Under the Directive, all the EU countries are required to use energy more efficiently at all stages of the energy chain from its production to its final consumption. EU countries were required to transpose the Directive's provisions into their national laws by 5 June 2014.

Compared to the last year, winters were relatively warm requiring less energy sources for heat generation. At the same time, the world economic crisis of 2008 affected significantly the economic situation of the Baltic States, and many representatives of manufacturing sector faced insolvency problems. The region in many aspects of economic development has still not recovered till the before-crisis level. Due to the stipulated RES targets, the share of gas consumption constantly shifts in favour of local green energy, but energy efficiency measures allow to save considerable part of generated energy.

The energy efficiency issue is very much linked to the issue of energy security, which is crucial for the Baltic States: the less natural gas is being consumed, the less it is imported.

5 Renewable Energy Sources

The primary aim of the policies addressing the renewable energy refers to the promotion of the environmental protection, particularly highlighted in the package of the European Union regulatory enactments in the field of environment and energy. The purpose of the initiative is to decrease the carbon emissions, as well as achieve by 2020 the stipulated in the *Directive 2009/28/EC of the European Parliament and of the Council of 23 April 2009 on the promotion of the use of energy from renewable sources and amending and subsequently repealing Directives 2001/77/EC and 2003/30/EC* target—20% share produced from RES energy in gross final energy consumption.

The Baltic Countries are quite different when it comes to the issue of renewable energy sources. These differences include both the starting positions in achieving the binding goals of the directive (see Fig. 5) and the actions taken. It should be noted that Latvia historically, already starting from late thirties of twentieth century,

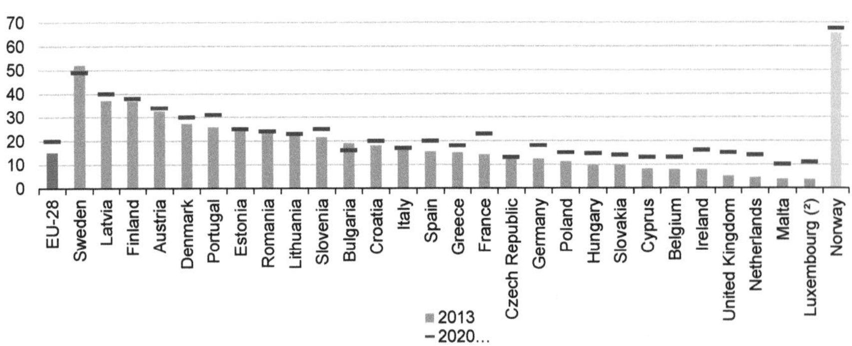

Fig. 5 The share of energy from RES in gross final energy consumption (2013 and 2020). *Source* Eurostat

Table 2 Production of RES by the source (technology)

	Primary production (thousand toe)		Share of total, 2013 (%)				
	2003	2013	Solar energy	Biomass and waste	Geothermal energy	Hydropower	Wind energy
EU28	104,094	191,961	5.5	64.2	3.1	16.6	10.5
Latvia	1728	2137	0.0	87.8	0.0	11.7	0.5
Lithuania	794	1288	0.3	92.1	0.1	3.5	4.0
Estonia	667	1122	0.0	95.7	0.0	0.2	4.1

Source Eurostat

has quite significant share of RES used in electricity production by Daugava cascade hydropower plants (48.8% in 2013 according to Eurostat).

Dependency on energy supplies makes the RES an essential part of the countries' energy policies. On the other hand, it makes the policies on the development of natural gas market more complicated to implement: in pursuit of gas market liberalization, the aspect of diminishing markets (also due to increasing share of RES in energy mix) may be an important argument of those opposing the market approaches.

Currently, the Baltic States do not cooperate on the issue of creating joint support mechanisms (or similar incentives) for the RES. Such cooperation would be definitely beneficial for the investors in the region ensuring economy of scale, as applying different support mechanisms within one commercial market (Nord Pool) distorts fair and transparent market conditions. However, at the same time, independency of policy making plays crucial role when it comes to possible competition for the investors. Despite this approach, the composition of the supported technologies is quite the same across the countries (see Table 2).

Nevertheless, the case of state support mechanism in Latvia proved to be the most discussed in the region, creating hot political debate and bringing three cases to the Constitutional Court.

To achieve the renewable energy target set out for Latvia, the state support mechanism was established for electricity generation from renewable sources or in high-efficiency cogeneration in a form of feed-in tariff. Aid beneficiaries according to the Latvian legislation are such merchants that produce electricity from RES, as well as in high efficient cogeneration by using biomass or biogas, or fossil energy resources (natural gas), and meet the qualification requirements for receiving the right to sell the produced electricity within the aid mechanism, receiving the fixed remuneration per kWh or for power stations above 4 MW—fixed amount of payment for capacity. According to the latest data at the disposal of the Ministry of Economics of Latvia, as on 1 January 2015, the number of the previous beneficiaries of the aid mechanism or power stations from which the electricity is purchased within the framework of the "mandatory procurement" or which receives the guaranteed payment for the installed electric capacity, is 389.

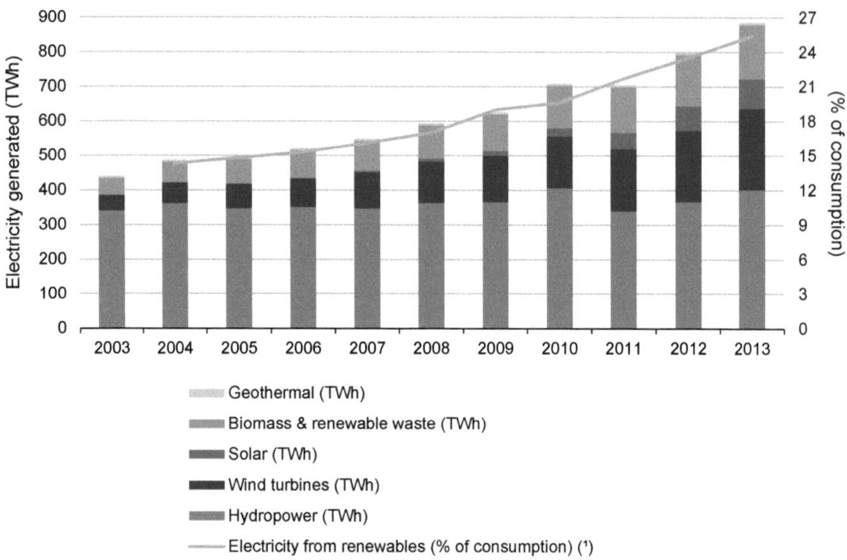

Fig. 6 The share of energy from RES in gross final energy consumption (2013 and 2020): European Union. *Source* Eurostat

Table 3 Payment made to the RES producers in Latvia (by technology) in 2014

	Produced amount of electricity, MWh	Amount paid to producers, MEUR	Support paid to producers (above the electricity market price), MEUR
RES	685,257	116.73	83.07
Biogas	335,539	62.39	45.64
Biomass	195,292	32.92	23.24
Wind (on shore)	87,786	9.42	5.36
Small-scale hydropower (till 5 MW)	66,641	12.00	8.83
CHP (fossil energy sources)	598,329	79.23	49.72
Capacity payment (RES)	–	5.16	
Capacity payment (CHP fossil energy sources)		107.78	
Total	1283,586	308.90	245.74

Source Ministry of Economics of the Republic of Latvia, 2015

Figure 6 quite clearly demonstrates that the support for the effective cogeneration or payments for the stability of the system (e.g. capacity payments) do not count as the country's or the European Union's effort in reaching RES goals (the other conclusion is that the biomass in the Baltic region is much more important than the other RES). However, in Latvia, capacity payment support was mixed with the traditional RES support. The gradually growing "mandatory procurement component" (the levy) cost in the total electricity bill of the final consumer has created general public non-acceptance of any support measures regarding the RES. The data in Table 3 actually show that the allocation of support was not entirely linked to the RES goals, but more to the overcoming the capacity and self-sufficiency concerns.

Given that Latvia has reached 37.1% by 2013, additional measures to achieve the renewable energy objective until 2020 would be required. Since the Latvian legislation requires that all electricity end-consumers are paying the levy proportionally to their final electricity consumption, reduced levy for energy-intensive companies will make it possible to maintain the competitiveness of businesses without affecting the renewable energy target. The cost of electricity is one of the main factors affecting the competitiveness of each economic sector, not only at national, regional, EU, but also on the international level. In particular, this aspect is applicable to those industries that have expanded their core business in energy-intensive sectors. The energy costs of the mentioned companies could reach up to 40% of the operational costs, which is much higher than similar companies have outside EU—for instance in Russia, USA, China, etc.

In recent years, the electricity cost in EU has been increased and individual EU member states have developed specific support mechanisms to support energy-intensive industries, at least to some extent to compensate the increase of the electricity cost. This existence of such support schemes had created the market distortion in export markets of Latvian energy-intensive companies putting those undertakings in a worse position.

The government decided to apply the support with the aim to reduce the negative impact on the competitiveness of the energy-intensive companies in export markets, where the electricity costs could be up to 2–3 times lower than that in Latvia. Therefore, the country filed to the EC the state aid scheme for energy-intensive industries in 2015 with no formal decision yet obtained (as of March 2016).

The state aid provisions proved to be complicated for the Baltic Countries: it took more than two years for Estonia to get approval of the state aid scheme and took more than 2.5 years for Latvia already (pre-notification started in late 2013). This is also caused by the turnover in state aid legislation, leading to the fact that for some period of time state aid schemes have not been considered as the state aid. In the Danish State aid case *Statsstøtte N 618/2003—Danmark Forlængelse af N 1037/1995 for visse kombinerede kraftvarmeværker*—when Commission considered that a scheme designed to support combined production of electricity and heat, did not qualify as State aid, grounded the EU member states' confidence about whether feed-in tariff is not to be considered State aid. In Danish circumstances, the Commission concluded that *"[t]here is therefore no loss of State funds" and that "the supplement which the power supply companies must pay above the market price of electricity [...], which is then passed on to final customers, does not constitute State aid within the meaning of Article 87(1) [EC]*".

This changed by the following court procedures and finally had been settled in the legislative proceedings—Guidelines on State aid for Environmental Protection and Energy 2014–2020. The final opinion of the European Commission is that the system of feed-in tariffs is to be considered as the state aid even if no direct public funds involved in the financing of the scheme.

Latvia's experience with application of support mechanisms is interesting due to introduction of subsidized electricity tax. In order to retain the aid in production of electricity by using RES or high-efficiency cogeneration plants and at the same time prevent substantial increase in levy, retaining it at the level of 2013 (2.69 EURcents/kWh), subsidized electricity tax has been introduced in Latvia. The subsidy payment provided to the RES produces has been set as the tax object. However, the maximum period of 4 years for tax application has been introduced. The subsidized electricity tax is one of the measures across the EU to reduce the payments to RES produces, which caused many discussions on predictability of European investment climate as well as on the necessity of current RES support mechanisms as lacking real market impact on the competition of the technologies.

5.1 Infrastructure and Interconnections

The built-in the last century approach in energy supply focuses on large-scale, centralized power production mainly using fossil fuels, as markets are not interconnected well enough. The main aim of the base-load-oriented system is to ensure a certain area with sufficient energy capacity for efficient development of a region without taking into account other neighbouring regions. The Energy Union strategy changes the approach and optimizes the available in the region capacities. Therewith, sufficient interconnections are the crucial precondition for optimal functioning of energy market.

Baltic electricity system has sufficient interconnections with European (North Europe) networks between Estonia and Finland, the direct current cables *Estlink 1* (Harku–Espoo; transmission capacity—350 MW) and *Estlink 2* (Puusi–Anttila; transmission capacity—650 MW) connecting to Estonia's and Finland's energy systems respectively.

At the end of 2015, the interconnection Lithuania–Sweden *NordBalt* (connecting Klaipeda, Lithuania and Nibro, Sweden) was finished, ensuring additional transmission capacity of 700 MW. This project is important for the integration of the Baltic States into the European systems enhancing energy supply security, as well as development of the regional market providing access to the rich with hydroenergy Nordic region.

Parallel to *NordBalt* at the end of 2015, the first stage of the Lithuania–Poland electricity interconnection *LitPolLink 1* with transmission capacity of 500 MW was finished. The network connects Kruonis hydroaccumulation power station (Lithuania) and Elk (Poland). After the implementation of the second stage, the transmission capacity of the interconnection will be increased till 1000 MW. Moreover, the discussions on necessity to build the *LitPolLink 2* project are also going on. This project could become an important element in case of change of

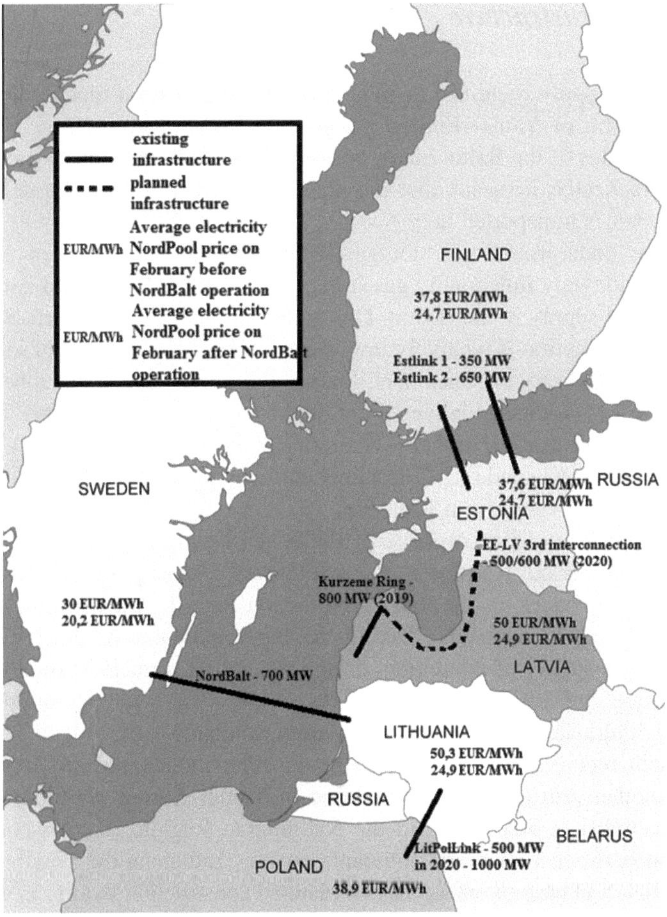

Fig. 7 Existing and planned electricity infrastructure projects. *Source* ENTSO-E

electricity system operation of the Baltic States to the synchronous mode with the continental Europe (Fig. 7).

To further strengthen Baltic transmission networks related to *NordBalt* project, the internal electricity line on western part of Latvia, so-called *Kurzeme Ring*, is being constructed. *Kurzeme Ring* is 330-kV electricity line planned to be finished in 2019.

The third Latvia–Estonia interconnection aiming to increase the transmission capacity between Latvia and Estonia by 500–600 MW is planned to be constructed by 2020. It will connect the Riga Thermal Electro station TEC2 and Kilingi-Nomme in Estonia. Both mentioned infrastructure projects are included in the third list of the EU projects of common interest according to the regulation No. 1316/2013 and have received financing from the European Union through Connecting Europe Facility financial instrument.

5.2 Gas Infrastructure

The main gas supply route to Latvia is 700-mm transmission pipeline Pskov-Riga which is an arm of Yamal–Europe gas pipeline coming from Russia. The transmission pipelines of the Baltic States are well developed and have capacity reserve of 40%. Practically, it means that the pipelines' capacity allows to deliver twice more gas as it is transported now.

Incukalns underground gas storage (IUGS) with total capacity of 2.3 bcm of active gas is the only functioning gas storage in the Baltic States, and it ensures the stability of gas supply in the region. During the summer period (April–September) when gas consumption is relatively low, gas is injected into IUGS and some part of it is delivered directly to consumers. During the winter period (October–March), gas from the gas storage is delivered not only to the Latvian customers, but also to the consumers in Estonia, Northern–Western Russia and in some cases to Lithuania. Since 2015 when Klaipeda LNG terminal started its operation, IUGS is also used for storing gas delivered from Klaipeda.

The maximum injection capacity of IUGS is 18 mcm/day. At the beginning of gas extraction period (October), the withdrawal capacity of IUGS can reach 30 mcm/day; however, at the end of extraction period (March), the withdrawal capacity falls till 10 mcm/day. Taking the big importance of the IUGS for the security on gas supply of the region, during the period from 1997 till 2013 for the modernization of IUGS 384.7 million EUR was spent, improving its injection/withdrawal daily capacity and safety standard.

Lithuania receives gas mainly by the pipeline interconnection from Belarus through another arm transmission pipeline of Yamal–Europe pipeline, as well as delivers in transit a part of gas to the Kaliningrad Region. A cross-border interconnection between Latvia and Lithuania ensures Lithuania the possibility to get gas from IUGS in case of emergency or planned construction works, as well as use storage services of IUGS.

The situation with security of gas supply in the region considerably improved with starting operation of Klaipeda LNG terminal in 2015. The terminal for the first time ensured alternative gas supply routes and sources to the Russian gas delivered by pipelines. Moreover, the capacity enhancement of Klaipeda–Kiemenai pipeline made it physically possible to ensure considerable gas supplies to all the three Baltic States.

The Estonian gas system is connected by transmission pipelines with Latvia and Russia. During the summer period, the gas to Estonian customers is delivered directly from Russia. However, during the winter season, the Estonian consumers are supplied mainly by gas from IUGS or from Russia through the gas metering station Värska. In particular cases, the gas is supplied from Russia through Narva.

Gas supply scenario in the Baltic States during the winter season in normal conditions is displayed in Fig. 8 and consists of the following key elements:

- Gas supply through Kotlovka (Belarus) GMS—19.75 mcm/day;
- Gas supply from Klaipeda LNG terminal—4.25 mcm/day;

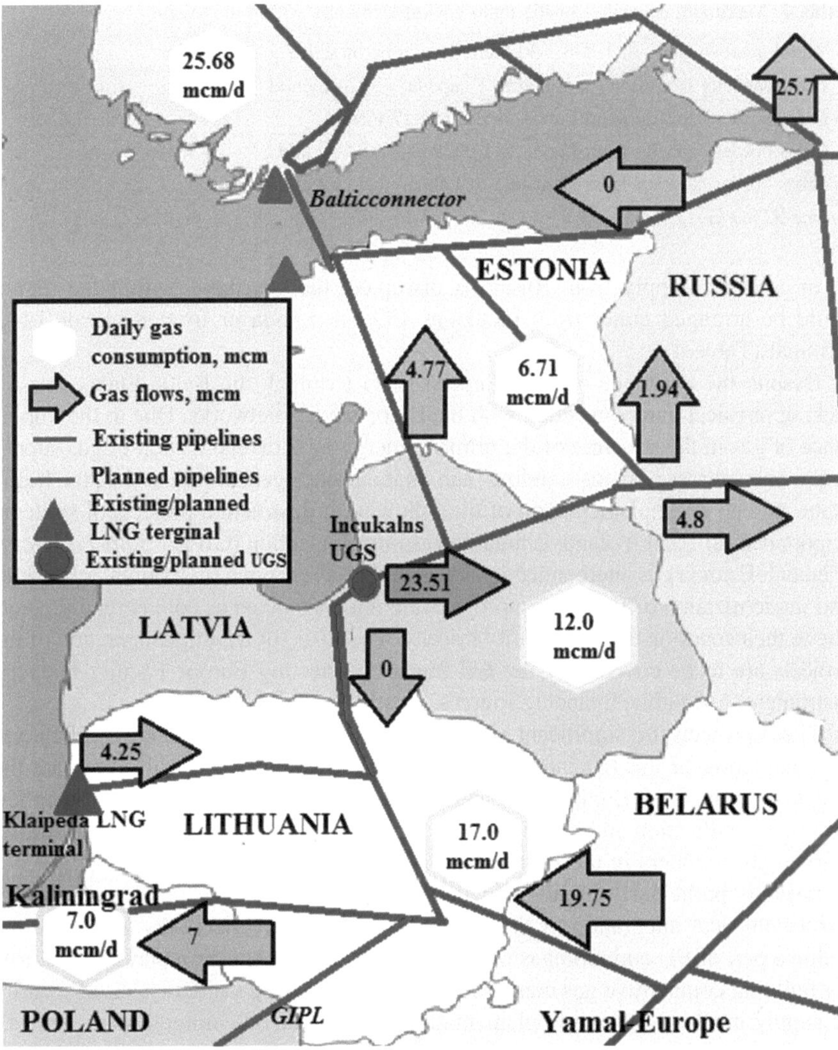

Fig. 8 Gas supply scenario in winter season in normal conditions. *Source* AS Latvijas Gaze

- Gas transit to Kaliningrad through Sakiai GMS—7.0 mcm/day;
- Gas supply from IUGS—23.5 mcm, including:
  - to Latvia—12.0 mcm/day;
  - to Karksi GMS—4.77 mcm/day;
  - to Verska GMS—1.94 mcm/day;
  - to Russia through Korneti GMS—4.8 mcm/day;
- Gas supply through Imatra (Finland) GMS—25.68 mcm/day.

Table 4 Maximum capacities of the main gas infrastructure within the region

– Withdrawal capacity of IUGS—30 mcm/d, volume of active gas—2.3 bcm.
– Entry capacity from LNG terminal in Klaipeda—10.24 mcm/d.
– Cross-border capacity from Latvia to Estonia—7 mcm/d.
– Cross-border capacity from Latvia to Lithuania—6.24 mcm/d.
– Cross-border capacity from Lithuania to Latvia—6.48 mcm/d.

Source ETSO-G

In case gas supply from Russia is disrupted, the gas flows within the region could be arranged either from Incukalns UGS in Latvia or from Klaipeda LNG terminal (Table 4).

Despite the existence of the Klaipeda LNG terminal, the Baltic States are still lacking physical interconnection with the European gas networks. Due to the importance of gas in the structure of the primary energy mix, diversification of gas supply routes and sources remains standing high on the Baltic agenda. Due to this, the Baltic States agreed on implementation of the following infrastructure projects of strategic importance till 2020: Poland–Lithuania gas interconnection (GIPL), *Balticconnector* (Finland–Estonia gas interconnection), regional LNG terminal (in Estonia or Finland) and modernization of IUGS. The projects are not considered as commercial projects due to their focus on energy security priority. The costs for the implementation of the projects are to be covered by the EU funds (Connecting Europe Facility financial instrument) and public financing sources, mostly through system tariffs.

These projects are significant elements for achieving the EU energy policy aim on completion of the EU internal energy market. The projects will eliminate the isolation of the three Baltic States and reduce dependency on a dominant supplier, ensure diversification and security of gas supplies, as well as enhance a competition among gas suppliers in the Baltic States gas market.

GIPL is particularly highlighted in the list of the strategic projects, as it will ensure physical integration of the Baltic States to the EU networks and make the Baltic a part of EU common gas market. The physical connection plays crucial role for building competitive gas market in the region, building solidarity-based security of supply mechanisms and implementation of harmonized common market rules.[2]

6 Market Drivers

6.1 Electricity Market

To facilitate competition on the Baltic States' electricity market and develop cross-border trade, the Baltic region joined Nord Pool Spot power exchange. The

[2]http://www.entsog.eu/public/uploads/files/publications/GRIPs/2014/GRIP_002_140514_BEMIP_2014-2023_main_low.pdf.

Table 5 Key indicators: Electricity (2014)

	Estonia[a]	Latvia[b]	Lithuania[c]
Number of companies representing at least 95% of net power generation	6	17	30
Number of main power generation companies	1	1	6
Market share of the largest power generation company (%)	87	89	25

Source Countries' reports to the EC
[a]https://ec.europa.eu/energy/sites/ener/files/documents/2014_countryreports_estonia.pdf
[b]https://ec.europa.eu/energy/sites/ener/files/documents/2014_countryreports_latvia.pdf
[c]https://ec.europa.eu/energy/sites/ener/files/documents/2014_countryreports_lithuania.pdf

Latvian trade zone started to operate in 2013, Lithuanian—in 2012, but Estonian—in 2010. Currently, all the trade transactions are realized only via Nord Pool.

Nord Pool Spot operates in Europe's leading power markets, offering both day-ahead and intraday trading to its members. Three hundred and eighty companies from 20 countries trade on Nord Pool Spot's markets in the Nordic and Baltic regions, and on UK market N2EX. In 2014, the group had a total turnover of 501 TWh traded power [nordpoolspot, annual report].

In the situation of liberal electricity market even for countries having sufficient generating capacities, quite often it is beneficial to buy energy at NordPool and to run own generation units in case there is a need to cover the balance. It is determined by the cheap electricity in the Nordic region generating power mainly by HPPs and NPPs. In 2014 the average electricity price in Latvia's and Lithuania's price zone was for 40% higher than that in Sweden and for 20–30% higher than that in Estonia. It is justified by the strong interconnection lines between Latvia and Lithuania and congestion on Latvia–Estonia border's electricity lines. After launching Lithuania–Sweden interconnection at the beginning of 2016, the prices in Latvia and Lithuania dropped by approximately 30% and became almost equal to Estonia.

The key indicators provided in Table 5 demonstrate the similarities of Latvia and Estonia in the structure of generation capacities (89 and 87%, respectively). On the other hand, Lithuania is quite different with the lowest market share of the largest power generation company (25%) and having six main power generating companies. However, the number of generating companies making 95% of the power market in Estonia is only 6, Latvia has 17, but Lithuania has 30 players.

6.2 *Gas Market*

Another factor acting as a strong argument for the Baltic States towards integration to the EU gas market is price difference. Traditionally, the lowest wholesale gas prices are in the trade platform of the UK *National Balancing Point* (NBP). The price-forming principles differ from country to country, and the price comparison could be done only for indicative purposes. Nevertheless, at time when in the first half of 2014 the gas price on NBP was 21.80 EUR/MWh, the gas prices on other trading platforms during equal period varied from 21.57 to 28.80 EUR/MWh.

Traditionally, the Baltic States belong to the group of countries in Europe with comparatively high price range from 27.51 to 30 EUR/MWh. During the referred period, the gas price in Latvia was 28.86 EUR/MWh, Lithuania—35.20, but Estonia—31.32 EUR/MWh. However, in contrast to the Baltic region, the HUB price in Poland was 22.22 EUR/MWh, in Germany 21.81 EUR/MWh and, for example, Holland—21.56 EUR/MWh.

The gas prices considerably influence the ability of local power generation installations to produce and offer to the market competitive electricity and heat products. In time of fierce competition and rapid development of RES using power generation technologies, the ability to ensure competitive gas price to the region will play an important role on a map of power base loads versus cross-border energy flows of the region.

The gas price in a particular region depends on global gas market trends, cross-border transmission congestions and competition level being directly affected by the legal framework, regulating the rules of the game in a certain country. To make the fifth EU freedom—free movement of energy—possible, the implementation of the third energy package is required. Historically, all the three Baltic states had vertically integrated gas companies ensuring transmission, distribution services (and storage in Latvia) and trade. Lithuania and Estonia have already completed gas market liberalization; however, Latvia intends to open the gas market on 3 April 2017. The gas opening deadline in Latvia respects the terms of privatization agreement signed between the strategic investors, when in 1997 the majority of the state-owned shares were sold to private entities. According to the Agreement, Latvijas Gāze JSC has exclusive rights on transmission, storage, distribution and trade until 2017. Nevertheless, the third-party access is ensured in all the three Baltic countries and market players can benefit from having access to the gas transmission systems and storage facility based on technical capabilities.

The key indicators provided in Table 6 demonstrate once again the similarities of Latvia and Estonia and quite different structure of the gas market in Lithuania. This difference is justified by the existence of the Klaipeda LNG terminal, establishing the real possibilities of alternative gas supplies.

The Klaipeda LNG terminal (Lithuania) has considerably widened the market possibilities of the region bringing gas supply alternatives in addition to Gazprom supplies. Consequently, in Lithuania gas is imported by five companies, including

Table 6 Key indicators: Natural gas (2014)

	Estonia	Latvia	Lithuania
Number of entities bringing natural gas into country	1	1	5
Number of main gas entities	2	1	4
Market share of the largest entity bringing natural gas (%)	86.5	100	69

Source Countries' reports to the EC

three gas supply companies (AB Lietuvos Dujos, UAB Dujotekana and UAB Haupas) and two major importers AB Achema and UAB Kauno Termofikacine Elektrine (Kaunas CHP), importing mainly for its own purposes.

In Estonia, practically two companies import natural gas—AS Eesti Gaas, which is a wholesaler on the market of Estonia, and AS Nitrofert, importing gas for own purposes. In Estonia are operating 24 distribution companies.

In Latvia, gas is imported by two companies—JSC Gazprom and less significant Itera Latvia delivering up to 25% of gas consumption in Latvia. Latvijas Gāze JSC remains the only distributor of gas on retail market.

7 System Security and Integration

Due to the historical reasons, the Baltic States have unique situation, as their electrical energy systems are operating in parallel synchronous mode not with the European systems, but with IPS/UPS (Integrated Power System/Unified Power System of Russia) region covering Russian energy system/Ukraine, Belarus, Kazakhstan, Kirgizstan, Azerbaijan, Georgian, Tajikistan, Moldavian and Mongolian integrated energy system, managed from Russia. The cross-border operation of electricity market between Baltic States, Russia and Belarus is regulated by the *BRELL* agreement signed between transmission system operators of Belarus, Russia, Estonia, Latvia and Lithuania (*BRELL*—abbreviation from Belarus, Russia, Estonia, Latvia, Lithuania). The Baltic States are responsible for the power system quality indicators within their territories. In the current situation, the Baltic electricity system is operated by Russia and directly depends on power generation process in the other third countries of IPS/UPS system. Practically, it means that launching or decommissioning of power installations within the system has impact on system security of the Baltic States. However, the Baltic region cannot affect the decisions on power generation units outside its borders. Moreover, the BRELL agreement restricts information exchange due to confidentiality, thereby fracturing the EU internal energy market.

Despite the fact that currently the system operation is technically stable, the functioning of electricity system remains strongly dependent on the third countries, which is the reason for the Baltic States to be called "energy island". Therewith, the Prime ministers of the three Baltic States have taken a decision on Baltic States power system synchronization with European networks. Currently, the synchronization project is high on the agenda of the region aiming at electricity market integration and increasing energy security.

The importance of synchronization project is highlighted in many political documents, such as European Council conclusions, European Energy Security Strategy and Declaration on Energy Security of Supply of the Baltic States signed in January 2015.

The synchronization with Continental Europe network is not a simple task, and it involves a range of infrastructure interconnections projects. Part of them is already under construction, attracting also the EU financing (*e.g. NordBalt or Lithuania–Sweden interconnection, Kurzeme Ring, LitpolLink1)*, and another part is on a way (e.g. the third Latvia–Estonia interconnection, *BalticCorridor*).

Since 1998, six different studies on de-synchronization of the Baltic States from IPS/UPS system have been implemented. The results of the studies show that the change of operation mode is technically possible, and the several solutions exist. The feasibility study "Integration of the Baltic States into the EU internal electricity market" was realized by the Swedish consulting company "Gothia Power", and the three Baltic transmission system operators "Augstsprieguma tīkls", "Elering" and "Litgrid" analysed the possible operation of the Baltic system with the continental Europe applying three different scenarios:

- Baltic States synchronous with IPS/UPS
- Baltic States and Kaliningrad synchronous with Continental Europe
- Baltic States synchronous with Continental Europe, but asynchronous with Kaliningrad.

The study also analysed the impact of large-scale power generation facility integration on the synchronization project. The findings showed that the proposed generation facility with capacity of 1350 MW (considering Visaginas Nuclear Power Plant project in Lithuania) is too big for the region with such small energy consumption as Baltic.

The synchronization project is also politically sensitive, as it influences operation of the electricity networks of the third countries. One of the questions to be answered is whether the Kaliningrad region should be included into the synchronous operation together with the Baltic States or not. The research showed that both the solutions are technically possible. The political decision should be taken first.

At the end of 2015, the Baltic States and the European Commission agreed to make another study on alternative solutions involving Joint Research Centre—the operation of the Baltic States in autonomous mode and synchronous operation of the Baltic States with the Nordic region. The study will cover the missing from the previous study elements, such as costs and benefits of the solutions, power generation costs (including CO_2 costs), system safety analysis, energy storage requirements and need for export/import, and operation interconnections.

The results of the study will be available at the end of 2016. The decision on the appropriate solution for de-synchronization from the IPS/UPS electricity system could be taken only after thorough analysis of the results of all the studies. The change of operation mode is planned for 2025.

8 Medium- and Long-Term Policy Vision

Due to the recent trends and ongoing infrastructure projects, the following factors will influence electricity prices in the region:

Infrastructure projects:

- Latvia–Estonia third electricity interconnection 500–600 MW (2020)
- Lithuania–Poland electricity interconnection LitPolLink, the second stage 500 MW (2020)
- Lithuania–Poland gas interconnection (2020)
- *Balticconnector* (2020)
- Regional LNG terminal (2020)

Generation capacities:

- Olkiluoto NPP in Finland 1600 MW (2018)
- Hanhikivi NPP in Finland 1200 MW (2024)
- Decommissioning of four NPPs in Southern Sweden (till 2020)
- Development of renewable energy sources

Generation costs:

- Change in prices for gas and oil;
- Changes due to CO_2 policy (2030).

Trends in the development of the electricity markets are and will be much linked with the integration to the European electricity networks, both in terms of infrastructure and pricing zones; however, the major challenge will be linked with the synchronization project:

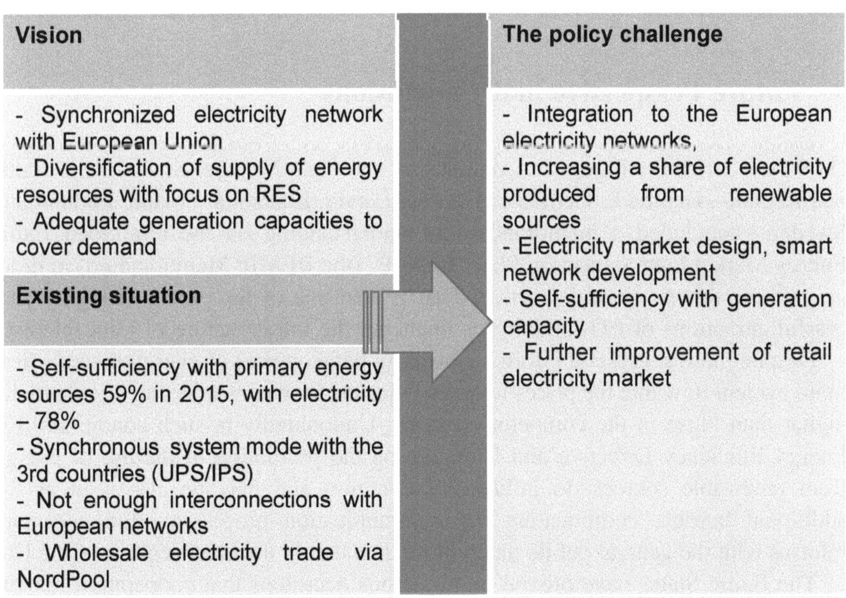

Taking into account the existing situation and the ongoing trends in the gas sector, the policy of the Baltic region is oriented on diversification of gas supply routes and sources, effective functioning of the gas market in the free market conditions and regional gas market development.

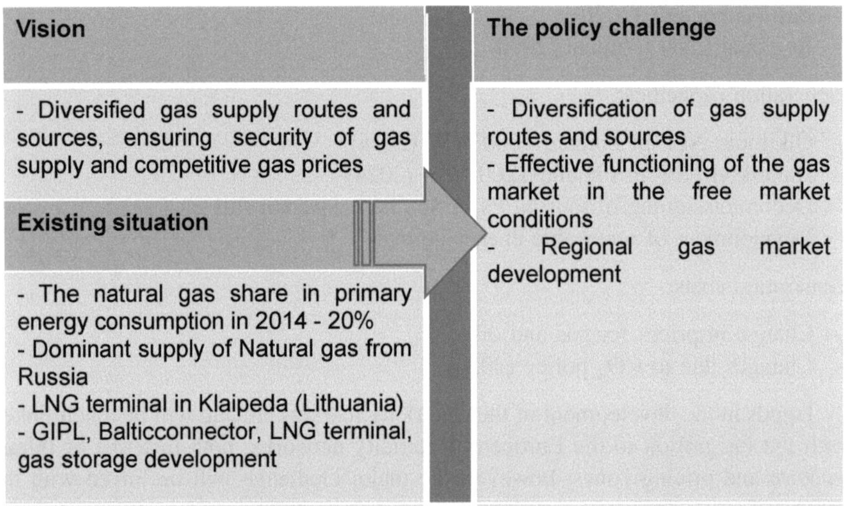

9 Future Perspective and Conclusions

On 9 June 2015, the European Commission and eight Member States of the Baltic Sea Region—Denmark, Germany, Estonia, Latvia, Lithuania, Poland, Finland and Sweden—concluded a memorandum of understanding on the reinforced Baltic Energy Market Interconnection Plan "BEMIP" (the BEMIP Memorandum) in order to continue this successful practice. That is also one of the proofs of highly successful attractions of EU funding for financing the infrastructure of joint interest.

Despite these successful efforts in ensuring development of energy systems, it is more evident now that the prices for energy for the final consumers are substantially higher than those in the competing regions. Undoubtedly in such conditions, EU Energy Efficiency Directive and Directive on the promotion of the use of energy from renewable sources do influence the region and may be also seen as the additional burden, complicating the communication process on the necessary reforms with the general public much more than that in the other regions of the EU.

The Baltic States have proved on numerous occasions that cooperation matters for smaller states both in terms of market integration and finance-intensive investment projects. However, a range of challenges are still ahead:

1. Synchronization project. It is not evident which scenario will be approved for implementation of the project, which is politically agreed to be completed by 2025. Extensive financial needs with simultaneous possible development of the Visaginas nuclear power plant make the project both technically and financially impossible to handle without extra support from the European Commission.
2. The electricity market is not yet fully completed: Lithuania still has the regulated process for the households; issue of the vulnerable consumers is to be addressed. Still the major issue is the construction of missing infrastructure lines given the fact that financing for these purposes has been granted.
3. The gas market is not fully liberalized: Latvia is on the way to catch the developments of the neighbouring countries; future model of operations of the market is an open issue. The countries should still agree on the gas pricing principles with the dependency on a single source of supplies remaining quite high.
4. Better cooperation in designing and implementation of the RES support schemes would have ensured more coherent investment policy in the sector. The future operations under BEMIP will be also linked to the Energy Efficiency Directive and RES since these are the integral parts of the whole energy policy concept. The issue of RES is linked to the issue of energy security in the Baltic States as these resources are indigenous energy resources of the region with the best chances to last for very long periods of time.
5. Generation capacity of the region has its weakness for the long-term perspective and should be addressed not obligatory by implementing a joint project, but also by coordinated implementation of various RES projects. Due to the safety concerns, the self-sufficiency of the region should be widened even if conflicting with the current price levels on the market.

References

1. European Commission. Commission staff working document. *Country Factsheet Estonia.* Accompanying the document. Communication from the commission to the European Parliament, the Council, the European Economic and Social Committee, the Committee of the Regions and the European Investment Bank. State of the Energy Union. 18.11.2015. SWD (2015)222 final.
2. European Commission. Commission staff working document. *Country Factsheet Latvia.* Accompanying the document. Communication from the commission to the European Parliament, the Council, the European Economic and Social Committee, the Committee of the Regions and the European Investment Bank. State of the Energy Union. 18.11.2015. SWD (2015)230 final.
3. European Commission. Commission staff working document. *Country Factsheet Lithuania.* Accompanying the document. Communication from the commission to the European Parliament, the Council, the European Economic and Social Committee, the Committee of the Regions and the European Investment Bank. State of the Energy Union. 18.11.2015. SWD (2015)231 final.

4. ENTSO-E. ENTSO-E Scenario Outlook & Adequacy Forecast (SO&AF) 2015. Available online: <https://www.entsoe.eu/Documents/SDC%20documents/SOAF/150630_SOAF_2015_publication_wcover.pdf>.
5. European Commission, Joint Research Centre, Purvins A., Fulli G., Covrig C., et al. The Baltic power system between East and West interconnections. Available online: <http://publications.jrc.ec.europa.eu/repository/bitstream/JRC100528/reqno_jrc100528_pdf.pdf> DOI:10.2790/411653
6. Single market progress report: Estonia. Available online: <https://ec.europa.eu/energy/sites/ener/files/documents/2014_countryreports_estonia.pdf>
7. Single market progress report: Latvia. Available online: <https://ec.europa.eu/energy/sites/ener/files/documents/2014_countryreports_latvia.pdf>
8. Judgment of the Court of 13 March 2001. - PreussenElektra AG v Schhleswag AG, in the presence of Windpark Reußenköge III GmbH and Land Schleswig-Holstein. - Reference for a preliminary ruling: Landgericht Kiel - Germany. - Electricity - Renewable sources of energy - National legislation requiring electricity supply undertakings to purchase electricity at minimum prices and apportioning the resulting costs between those undertakings and upstream network operators - State aid - Compatibility with the free movement of goods. - Case C-379/98. *European Court reports* 2001, Page I-02099
9. Single market progress report: Lithuania. Available online: <https://ec.europa.eu/energy/sites/ener/files/documents/2014_countryreports_lithuania.pdf>
10. Communication from the Commission—Guidelines on State aid for environmental protection and energy 2014-2020. Official *Journal of the European Union* C 200/1
11. ENTSO-G. Baltic Energy Market Interconnection Plan GRIP. Available online: <http://www.entsog.eu/public/uploads/files/publications/GRIPs/2014/GRIP_002_140514_BEMIP_2014–2023_main_low.pdf>
12. Eurostat data base. Available online: http://ec.europa.eu/eurostat/data/database;
13. Commission staff working document. European Union Strategy for the Baltic Sea Region. Action Plan. Brussels, 10.9.2015. SWD(2015) 177 final, COM(2009) 248
14. Gothia Power. Executive summary of Project "Feasibility study on the interconnection variants for the integration of the Baltic States to the EU internal electricity market". Available online: http://www.ast.lv/files/ast_files/files/documents/Executive%20summary%20of%20FIS-BIS%20project.pdf

Active Distribution Networks Operation Within a Distribution Market Environment

Geev Mokryani

Abstract This chapter proposes a novel method for the operation of active distribution networks within a distribution market environment taking into account multi-configuration of wind turbines. Multi-configuration multi-scenario market-based optimal power flow is used to maximise the social welfare considering uncertainties related to wind speed and load demand. Scenario-based approach is used to model the uncertainties. The method assesses the impact of multiple wind turbine configurations on the amount of wind power that can be injected into the grid and the distribution-locational marginal prices throughout the network. The effectiveness of the proposed method is demonstrated with 16-bus UK generic distribution system.

Keywords Wind power · Active network management · Social welfare · Market-based optimal power flow · Distribution network operators · Distribution-locational marginal prices

1 Introduction

1.1 Motivation and Approach

Renewable energy sources (RES) integration into distribution networks introduces several technical and economic challenges to distribution network operators (DNOs). The optimal allocation of distributed generators (DGs) may define several benefits including positive capacity margin, losses reduction, energy savings, voltage control, ancillary services, transmission and distribution capacity deferral and higher power quality. The impacts of RES connection on the network rely on different parameters such as: size, type and location of the new connections,

G. Mokryani (✉)
School of Electrical Engineering and Computer Science, University of Bradford, Bradford BD7 1 DP, UK
e-mail: g.mokryani@bradford.ac.uk

B. Azzopardi (ed.), *Sustainable Development in Energy Systems*,
DOI 10.1007/978-3-319-54808-1_6

the density of installations and proximity to the load, the pattern and timing of output, the state of the network and the overall amount of capacity [1–4].

Renewable energy sources are supposed to develop the design and operation of distribution networks. On the other hand, emerging active network management (ANM) schemes have proved to be advantageous for DNOs, compared to passive network management [5]. ANM schemes can increase the operation of the assets of network that allow the distribution networks to accommodate more distributed generators (DGs) within the existing infrastructure. The main ANM schemes include coordinated voltage control (CVC) of on load tap changers (OLTCs) and voltage regulators, adaptive power factor control (PFC) of DGs and energy curtailment [6, 7].

This chapter provides a novel approach for DNOs to assess the amount of wind power that can be injected into a distribution network considering uncertainties related to the stochastic variations of wind power generation and load demand, multiple WT configurations and ANM schemes including CVC and PFC. The method also characterises the impact of the above-mentioned factors on the distribution-locational marginal prices (D-LMPs). Multi-configuration multi-scenario market-based optimal power flow (MMMOPF) is utilised to maximise the social welfare (SW) considering above-mentioned uncertainties. A distribution market model is presented here under a distribution market structure based on pool and bilateral contracts within DNO's control area [8]. Here, the DNO is defined as the market operator of the distribution market, which determines the price estimation and the optimisation process for the hourly acquisition of active power [9].

Many works have been carried out about the power system operation. A method, based on bi-level modelling, for the optimal planning of incentives to promote wind power investment in transmission networks is described in [10]. A procedure to determine the optimal allocation of wind capacity in transmission networks considering security constraints is proposed in [11]. A methodology for wind power investment in transmission networks from the point of view of the network operator is proposed in [12]. A stochastic mathematical program with equilibrium constraints (MPEC) is used to minimise consumer payments. In [13], a method is proposed to identify the wind power plants to be built in transmission networks with the aim of maximising the profit of investors. The model is based on a stochastic MPEC model within a pool-based electricity market. In [14], wind power investment within a market environment in the context of a feed-in tariff scheme is investigated. The optimal wind power investment plan is selected by using mixed-integer nonlinear programming and a screening and ranking technique.

Several studies have been reported on the benefits of ANM and its applications. Some of them revealed implementations, and experiences of ANM [15, 16] and online ANM application [17, 18]. The cost–benefit analysis of investments and operation costs for various combinations of ANM schemes and techno-economic evaluation are studied in [19–21] and compared with passive network management scheme.

To the best of author's knowledge, no stochastic method for the operation of active distribution networks within a distribution market environment considering multiple WT configurations has been reported in the literature. In the above-mentioned references, the authors have not addressed the impact on the overall DG penetration level when one or more existing DGs are absent. Moreover, the presence of a distribution market environment has not been addressed in the above-mentioned studies. One of the main contributions of this chapter is proposing a novel MMMOPF-based approach, which takes into account the operational status of WTs, and assesses the dispatched active power of WTs considering different multi-configurations within the DNO acquisition market environment which has not been addressed so far. It provides detailed analysis and results on how multiple WT configurations could impact the amount of wind power that can be injected into active distribution networks as well as the D-LMPs throughout the network.

The rest of this chapter is organised as follows: Multi-WT configurations and uncertainty modelling are discussed in Sects. 2 and 3, respectively. Problem formulation is described in Sect. 4. Section 5 presents the 16-bus UK generic distribution system (UKGDS) and simulation results. Conclusions are presented in Sect. 6.

2 Multi-Wind Turbine Configurations

In this chapter, the MMMOPF method aims to incorporate multi-WT configurations, which is defined as the operational status of WTs, and is chosen based on the DNO's decisions. The total number of all possible multi-configurations for any number of WTs can be expressed as follows:

$$1 \leq \mathrm{NC} \leq (2^{\mathrm{NW}} - 1) \tag{1}$$

where NC and NW are the number of configurations and WTs, respectively.

The total configurations are referred as the number of multi-WT configurations. For example, if a system has three WTs, there will be up to seven possible multi-WT configurations for the DNOs to choose. A binary parameter is defined to represent the operational status of WTs at ith bus for configuration c. The operational status of each WT and all WTs are described in (2) and (3), respectively.

$$\beta_{i,c} = \begin{cases} 1, & \text{if a wind turbine at } i\text{th bus is operating} \\ 0 & \text{otherwise} \end{cases} \tag{2}$$

$$\beta = \begin{bmatrix} \beta_{1,w_1} & \beta_{1,w_2} & \cdots & \beta_{1,w_N} \\ \beta_{2,w_1} & \beta_{2,w_2} & \cdots & \beta_{2,w_N} \\ \vdots & \vdots & \ddots & \vdots \\ \beta_{c,w_1} & \beta_{c,w_2} & \cdots & \beta_{c,w_N} \end{bmatrix}_{(\mathrm{NC} \times \mathrm{NW})} \tag{3}$$

In the proposed method, there is capacity constraint for WTs according to its operational status for each configuration which is described as follows:

$$0 \leq P^{w}_{i,c} \leq \beta_{i,c} P^{w}_{i,c} \tag{4}$$

$$P^{w}_{i,c} = \begin{cases} 0 \leq P^{w}_{i,c} \leq P^{w,\max}_{i,c}, & \forall \quad \beta_{i,c} = 1 \\ 0, & \forall \quad \beta_{i,c} = 0 \end{cases} \tag{5}$$

where β *is the* operational status of each WT. $\beta_{i,c}$ is operational status of WTs at bus i and configuration c. NC and NW are, respectively, number of WTs and configurations. $P^{w}_{i,c}$ is the active power generated by WTs at bus i and configuration c.

3 Uncertainty Modelling

Using the technique described in [22, 23], the PDF of wind speed is divided into several intervals, and the probability of falling into each interval is calculated. A mean value is also assigned for each interval as an indicator of the corresponding interval. Load demands are also modelled by a normal PDF in which the mean and variance are known. It is assumed that the load demand and wind power generation scenarios are independent, therefore the scenarios are combined to construct the whole set of scenarios as follows:

$$\pi_s = \pi_D \times \pi_w \tag{6}$$

where π_D, π_w and π_s are the probabilities of Dth load, wth wind and whole set of scenarios, respectively.

4 Problem Formulation

4.1 DNO Acquisition Market Formulation

A DNO energy acquisition market model called the DNO acquisition market is presented here under a distribution market structure based on pool and bilateral contracts. The DNO is defined as the market operator of the acquisition market, which determines the price estimation and the optimisation process for the acquisition of active power. Loads and WTs send active power offers and bids to the DNO acquisition market in the form of blocks for each hour. More details about this can be found in [24–28]. Under the assumed DNO acquisition market, the market clearing quantity and price are determined by maximising the SW considering network constraints. The MMMOPF is formulated as follows:

$$\text{Maximize SW} = \sum_{s=1}^{NS} \pi_s \left\{ \sum_{q=1}^{NQ} \sum_{i=1}^{NB} \sum_{s=1}^{NS} \sum_{y=1}^{NY} C_{i,q}^{D} P_{i,q,s}^{D} - \sum_{t=1}^{NT} \sum_{i=1}^{NB} \sum_{s=1}^{NS} \sum_{c=1}^{NC} C_{i,t}^{w} P_{i,t,s,c}^{w} \right\} - \sum_{t=1}^{NT} \sum_{i=1}^{NB} \sum_{c=1}^{NC} C_{i,t}^{G} P_{i,t,c}^{G} \tag{7}$$

subject to

(a) *Equality Constraints: Active and Reactive Power Balance at Each Bus*

$$\left(\sum_{t=1}^{NT} \sum_{i=1}^{NG} P_{i,t,c}^{G} + \sum_{t=1}^{NT} \sum_{i=1}^{NB} P_{i,t,s,c}^{w} - \sum_{q=1}^{NQ} \sum_{i=1}^{NB} P_{i,q,s}^{D} \right) = \sum_{j=1}^{NB} V_{i,s,c} V_{j,s,c} T_{ij} G_{ij} \cos(\delta_{i,s,c} - \delta_{j,s,c}) + B_{ij} \sin(\delta_{i,s,c} - \delta_{i,s,c}) \tag{8}$$

$$\left(\sum_{t=1}^{NT} \sum_{i=1}^{NG} Q_{i,t,c}^{G} + \sum_{t=1}^{NT} \sum_{i=1}^{NB} Q_{i,t,s,c}^{w} - \sum_{q=1}^{NQ} \sum_{i=1}^{NB} Q_{i,q,s}^{D} \right) = \sum_{j=1}^{NB} V_{i,s,c} V_{j,s,c} T_{ij} G_{ij} \sin(\delta_{i,s,c} - \delta_{j,s,c}) - B_{ij} \cos(\delta_{i,s,c} - \delta_{i,s,c}) \tag{9}$$

where $(P/Q)_{i,q,s}^{D}$ is the active/reactive consumption of loads at bus i, block q, scenario s and configuration c. $(P/Q)_{i,t,s,c}^{w}$ is the active/reactive power generated by wind turbines at bus i, block t, scenario s and configuration c. $(P/Q)_{i,t,c}^{G}$ are the active/reactive power at slack bus, block t and configuration c. $C_{i,q}^{D}$ is the price for the energy bid q at bus i submitted by load D. $C_{i,t}^{w}$ is the price for the energy selling t at bus i by WT w. $C_{i,t}^{G}$ is the price for the energy selling t at slack bus. $V_{i,s,c}/\delta_{i,s,c}$ are, respectively, the voltage/voltage angle at bus i, scenario s and configuration c. G_{ij}/B_{ij} are the real/imaginary part of the element in the admittance matrix corresponding to the ith row and jth column.

(b) *Inequality Constraints*

- Branch flow constraints

$$\sqrt{\left(G_{ij}^2 + B_{ij}^2\right)\left(V_{i,s,c,y}^2 + \frac{V_{j,s,c,y}^2}{T_{ij}^2} - \frac{2V_{i,s,c,y}V_{j,s,c,y}\cos(\delta_{i,s,c,y} - \delta_{i,s,c,y})}{T_{ij}}\right)} \le I_{ij}^{\max} \tag{10}$$

where $I_{ij}^{\max}$ is maximum current flow of wires.

– Voltage limits at each bus

$$V_i^{\min} \le V_{i,s,c} \le V_i^{\max} \tag{11}$$

$$\delta_i^{\min} \le \delta_{i,s,c} \le \delta_i^{\max} \tag{12}$$

where $V_i^{\min}/V_i^{\max}$ and $\delta_i^{\min}/\delta_i^{\max}$ are min/max values of voltage and voltage angle at each bus, respectively.

– WTs generation constraint

$$0 \le P_{i,t,s,c}^{w} \le \gamma_{i,s,c}^{w} \times P_{i,\text{rated}}^{w} \tag{13}$$

$$Q_i^{w,\min} \le Q_{i,t,s,c}^{w} \le Q_i^{w,\max} \tag{14}$$

where $\gamma_{i,s,c}^{w}$ is the percentage of active power generated by WTs at scenario s and configuration c. $P_{i,\text{rated}}^{w}$ is WTs rated active power. $Q_i^{w,\min}/Q_i^{w,\max}$ are min/max values of reactive power of WTs.

– Capacity constraints at slack bus

$$P_i^{G,\min} \le P_{i,t,c}^{G} \le P_i^{G,\max} \tag{15}$$

$$Q_i^{G,\min} \le Q_{i,t,c}^{G} \le Q_i^{G,\max} \tag{16}$$

where $P_i^{G,\min}/P_i^{G,\max}$ and $Q_i^{G,\min}/Q_i^{G,\max}$ are min/max values of active and reactive power at slack bus, respectively.

– **ANM Schemes** [29–32]

(1) *Coordinated Voltage Control*

$$T_{ij}^{\min} \le T_{ij} \le T_{ij}^{\max} \tag{17}$$

where T_{ij} is the tap magnitude of OLTC. $T_{ij}^{\min}/T_{ij}^{\max}$ are the min/max values it can assume.

(2) *Adaptive Power Factor Control*

$$\phi_i^{w,\min} \le \phi_{i,s,c}^{w} \le \phi_i^{w,\max} \tag{18}$$

where $\phi_{i,s,c}^{w}$ is the power factor angle of WTs at bus i, scenario s and configuration c. $\phi_i^{w,\min}/\phi_i^{w,\max}$ are the min/max values they can assume.

5 Case Study and Simulation Results

In this section, the distribution system used to test the proposed method is described. The following analyses are based on 33 kV 16-bus rural weakly meshed UKGDS whose data are available in [33]. The single-line diagram of the distribution system is shown in Fig. 1.

The feeders are supplied by two identical 30-MVA 132/33 kV transformers. Two OLTCs, allocated between buses 1 and 2 has a target voltage of 1.05 p.u. at the secondary. A voltage regulator (VR) is located between buses 8 and 9, with the latter having a target voltage of 1.03 p.u..Voltage limits are taken to be ± 6% of nominal value, i.e. $V_{\min} = 0.94$ and $V_{\max} = 1.06$ p.u. and the power factor of WFs ranges from 0.95 leading to 0.95 lagging. In this paper, it is assumed that buses 5, 7 and 9 are three possible WFs locations, but it is notable that the selection of possible WFs' locations relies on non-technical factors such as legal requirements, space/land availability and other amenities. Three states for loads and three states for wind power generation are considered, respectively, by using normal and Weibull PDFs. By incorporating these scenarios, as explained in Sect. 3, nine combined wind load scenarios are obtained as given in Table 1.

Three 15 MW wind farms (WFs) are installed at buses 5, 7 and 9. Each of them is composed of 5 × 3 MW WTs. It is assumed that maximum four WFs can be allocated at each candidate bus. For each scenario and configuration, this is represented by four equal blocks in the WF's offer with the same price. The offer price of WFs is calculated using the method described in [24–28] which is equal to 27.84 £/MWh for each WF. The offer price at slack bus is assumed to be 150 £/MWh. Regarding the bids of loads, it is assumed that there are two blocks for each load as presented in Table 2 at maximum load. Table 3 presents all the possible multi-WT configurations for the three WFs locations using (1).

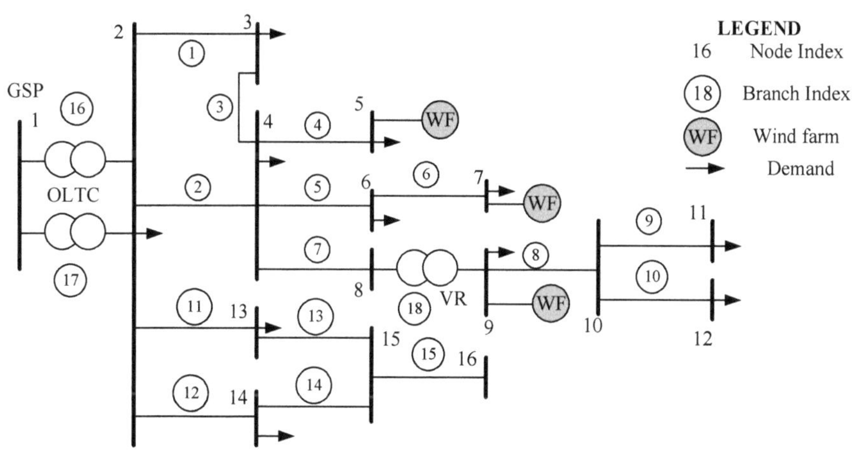

Fig. 1 16-bus UKGDS with candidate locations for WFs

Table 1 Combined wind and load scenarios and corresponding probabilities

Scenario #	Load (%)	Wind (%)	π_s
s_1	98	100	0.015
s_2	100	100	0.07
s_3	102	100	0.015
s_4	98	85	0.12
s_5	100	85	0.56
s_6	102	85	0.12
s_7	98	75	0.015
s_8	100	75	0.07
s_9	102	75	0.015

Table 2 Bid quantity and price of loads

Bus No.	Quantity (MW)		Price (£/MWh)	
	Block 1	Block 2	Block 1	Block 2
2	4.00	1.50	250	200
3	1.00	0.99	250	200
4	0.06	0.06	200	200
5	10.00	9.20	150	150
6	1.06	0.90	300	200
7	0.30	0.25	350	150
9	1.05	0.95	300	200
10	1.50	1.20	225	175
11	2.15	0.70	150	150
12	0.42	0.39	150	125
13	0.51	0.50	100	100
14	0.48	0.10	200	200

Table 3 Description of multi-WT configurations

Multi-configurations	WT status/location		
	Bus 5	Bus 7	Bus 9
1	1	0	0
2	0	1	0
3	0	0	1
4	1	1	0
5	1	0	1
6	0	1	1
7	1	1	1

Figure 2 shows the total dispatched active power of WFs for each configuration. Configuration 2 has the lowest dispatched active power compared to other configurations, while configurations 1 and 3 (i.e. one WF at buses 5 and 9, respectively) have the higher dispatched active power compared to that at bus 7. This is mainly due to the higher bid price and lower bid quantity, as presented in Table 2,

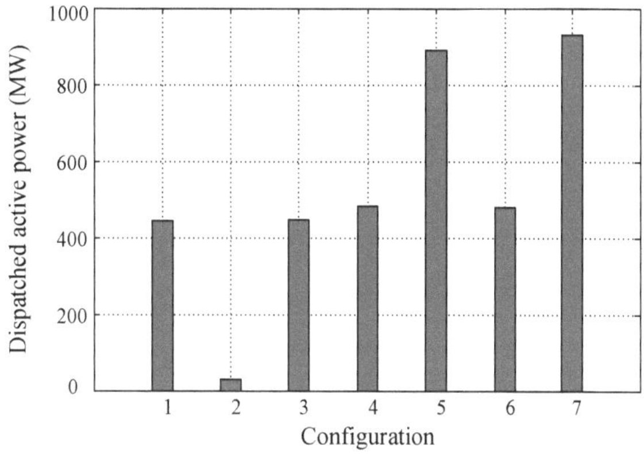

Fig. 2 Total dispatched active power for each configuration

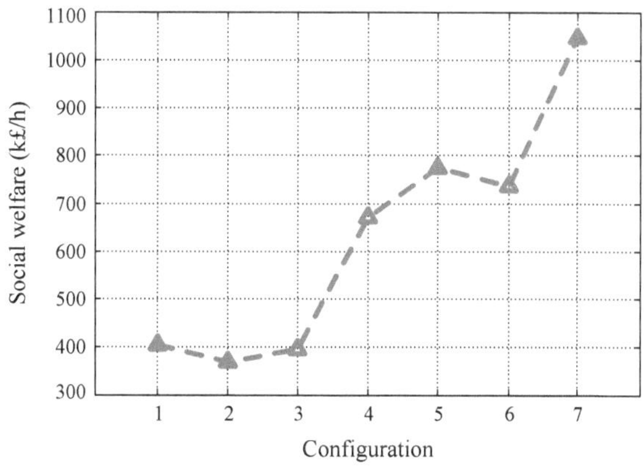

Fig. 3 Total social welfare for each configuration

and voltage constraints at bus 7 compared to those at buses 5 and 9 as well as the thermal limits of the lines connecting the buses. Configuration 5 has the higher dispatched active power compared to that in configurations 4 and 6. This is because of the higher bid quantity and lower bid price at buses 5 and 9 compared to those in configurations 4 and 6. It is seen that the dispatched active power in configuration 4 is almost equal to that in configuration 3. It is evident that configuration 7 with three WFs installed at candidate buses has the highest dispatched active power compared to that in other configurations.

The total SW for each configuration is shown in Fig. 3. It is seen that configurations 2 and 7, respectively, have the lowest and highest values of SW compared to others. This is mainly because of the lowest and highest dispatched active power at these configurations, respectively, as WTs allocation allows increasing the SW. Figure 4 shows the D-LMPs at candidate buses for each configuration. It is observed that at bus 7 and configuration 2, the D-LMP has the highest value. This is mainly because of the lowest dispatched active power at this bus and configuration. It is seen that the D-LMP in configuration 3 at bus 9 is about 500 £/MWh,

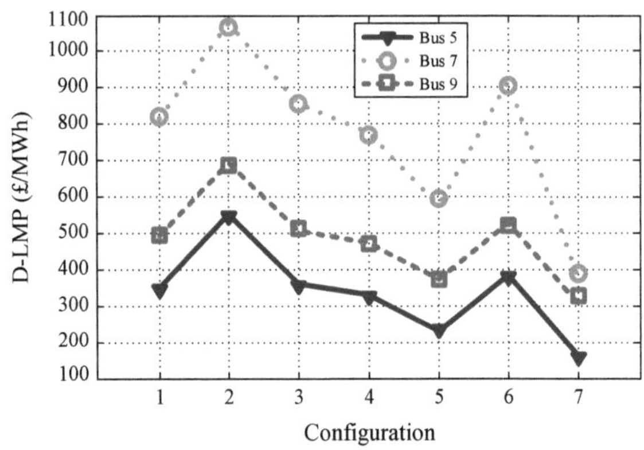

Fig. 4 D-LMP for each configuration at each bus

while this value in configuration 4 is about 480 £/MWh. This is due to the almost equal dispatched active power at these configurations as shown in Fig. 2. Therefore, configurations 1, 3 and 5 are more economical than configurations 2, 4 and 6. Thus, the method can be used as a useful tool for DNOs to install WTs at more advantageous locations in terms of consumers' benefits and cost reduction.

6 Conclusions

This chapter proposes a stochastic method for the operation of active distribution networks within a distribution market environment. MMMOPF is used to maximise SW considering uncertainties related to wind speed and load demand. It is revealed that the multi-WT configurations under ANM schemes could increase the potential of wind power penetration at certain locations and consequently decrease D-LMPs throughout the network. The proposed method can be used as a tool for DNOs to assess the impact of wind power penetration on a given network in terms of technical and economic effects.

References

1. R. Banosa, F. Manzano-Agugliaro, F.G. Montoya, C. Gila, A. Alcayde, J. Gómez, "Optimization methods applied to renewable and sustainable energy: A review", *Renewable and Sustainable Energy Reviews*, vol. 15, no. 4, pp. 1753–1766, 2011.
2. D. Kirschen and G. Strbac, *Fundamentals of Power System Economics*. New York: Wiley, 2004.
3. A. Piccolo, P. Siano, "Evaluating the impact of network investment deferral on distributed generation expansion", *IEEE Trans. Power Syst.*, vol. 24, no. 3, pp. 1559–1567, 2009.

4. G. Mokryani, P. Siano, A. Piccolo and Zhe Chen, "Improving Fault Ride-Through Capability of Variable Speed Wind Turbines in Distribution Networks", *IEEE Systems J.*, vol. 7, no. 4, pp. 713–722, 2013.
5. P. Djapic, C. Ramsay, D. Pudjianto, G. Strbac, J. Mutale, N. Jenkins, and R. Allan, "Taking an active approach," *IEEE Power Energy Mag.*, vol. 5, no. 4, pp. 68–77, 2007.
6. S. N. Liew and G. Strbac, "Maximising penetration of wind generation in existing distribution networks," *IEE Gener. Transm. Distrib.*, vol. 149, no. 3, pp. 256–262, 2002.
7. P. Siano, P. Chen, Z. Chen, and A. Piccolo, "Evaluating maximum wind energy exploitation in active distribution networks," *IET Gener. Transm. Distrib.*, vol. 4, no. 5, pp. 598–608, 2010.
8. G. Mokryani, P. Siano, "Strategic placement of distribution network operator owned wind turbines by using market-based optimal power flow," *IET Gener. Transm. Distrib.*, vol. 8, no. 2, pp. 281–289, 2014.
9. R. Palma-Behnke, J. L. A. Cerda, L. Vargas, and A. Jofre, "A distribution company energy acquisition market model with the integration of distribution generation and load curtailment options," *IEEE Trans. Power Syst.*, vol. 20, no. 4, pp. 1718–1727, 2005.
10. Y. Zhou, L. Wang, and J. D. McCalley, "Designing effective and efficient incentive policies for renewable energy in generation expansion planning," *App. Energy*, vol. 88, no. 6, pp. 2201–2209, 2011.
11. D. J. Burke and M. O'Malley, "Maximizing firm wind connection to security constrained transmission networks," *IEEE Trans. Power Syst.*, vol. 25, no. 2, pp. 749–759, May 2010.
12. L. Baringo, A. J. Conejo, "Transmission and Wind Power Investment", *IEEE Trans. Power Syst.*, vol. 27, no. 2, pp. 885–893, 2012.
13. L. Baringo, A. J. Conejo, "Wind power investment within a market environment," *Appl. Energy*, vol. 88, no. 9, pp. 3239–3247, 2011.
14. C. Kongman and S. Nuchprayoon, "Feed-in tariff scheme for promoting wind energy generation," in Proc. *IEEE PowerTech*, Bucharest, Romania, 2009.
15. R. A. F. Currie, G. W. Ault, R. W. Fordyce, D. F. MacLeman, M. Smith, and J. R. McDonald, "Actively managing wind farm power output," *IEEE Trans. Power Syst.*, vol. 23, no. 3, pp. 1523–1524, 2008.
16. O. Samuelsson, S. Repo, R. Jessler, J. Aho, M. Karenlampi, and A. Malmquist, "Active distribution network—Demonstration project ADINE," in *Proc. IEEE PES Innovative Smart Grid Technologies Conf. Europe (ISGT Europe)*, Oct. 2010.
17. M. J. Dolan *et al.*, "Distribution power flow management utilizing an online optimal power flow technique," *IEEE Trans. Power Syst.*, vol. 27, no. 2, pp. 790–799, 2012.
18. F. Pilo, G. Pisano, and G. G. Soma, "Optimal coordination of energy resources with a two-stage online active management," *IEEE Trans. Ind. Electron.*, vol. 58, no. 10, pp. 4526–4537, 2011.
19. A. Shafiu, T. Bopp, I. Chilvers, and G. Strbac, "Active management and protection of distribution networks with distributed generation", in *Proc. IEEE Power Eng. Soc. General Meeting*, 2004.
20. R. Hidalgo, C. Abbey, and G. Joos, "Technical and economic assessment of active distribution network technologies," in *Proc. IEEE Power Energy Soc. General Meeting*, 2011.
21. Z. Hu and F. Li, "Cost-benefit analyses of active distribution network management, part I: Annual benefit analysis," *IEEE Trans. Smart Grid*, vol. 3, no. 3, pp. 1067–1074, Sep. 2012.
22. Y. M. Atwa and E. F. El-Saadany, "Probabilistic approach for optimal allocation of wind-based distributed generation in distribution systems" *IET Renew. Power Gener.*, vol. 5, no. 1, pp. 79–88, 2011.
23. A. Rabiee, A. Soroudi, B. Mohammadi-Ivatloo, M. Parniani, "Corrective voltage control scheme considering demand response and stochastic wind power", *IEEE Trans. Power Syst.*, vol.29, no.6, pp. 2965–2973, 2014.
24. P. Siano, G. Mokryani, "Evaluating the benefits of optimal allocation of wind turbines for distribution network operators", *IEEE Syst. J.*, vol.9, no.2, pp. 629–638, 2015.

25. G. Mokryani, P. Siano, "Optimal wind turbines placement within a distribution market environment", *Applied Soft Computing*, vol.13, no.10, pp. 4038–4046, 2013.
26. G. Mokryani, P. Siano, "Combined Monte Carlo simulation and OPF for wind turbines integration into distribution networks", *Electr. Power Syst. Res.*, vol.103, pp. 37–48, 2013.
27. G. Mokryani, P. Siano, A. Piccolo, "http://link.springer.com/article/10.1007/s12652-012-0163-6" "Optimal allocation of wind turbines in microgrids by using genetic algorithm", *Journal of Ambient Intelligence and Humanized Computing*, vol. 4, no. 6, pp. 613–619, 2013.
28. P. Siano, G. Mokryani, "Assessing wind turbines placement in a distribution market environment by using particle swarm optimization", *IEEE Trans. Power Syst.*, vol.28, no.4, pp. 3852–3864, 2013.
29. N. Acharya, P. Mahat, and N. Mithulananthan, "An analytical approach for DG allocation in primary distribution network," *Int. J. Elect. Power & Energy Syst.*, vol. 28, no. 10, pp. 669–678, 2006.
30. M. Tsili, S. Papthanssiou, "A review of grid code technical requirements for wind farms, *IET Renew. Power Gener.*, vol.3, no.3, pp. 308–332, 2009.
31. G. Mokryani, A. Majumdar, B.C. Pal, "A probabilistic method for the operation of three-phase unbalanced active distribution networks", *IET Renew. Power Gener*, vol.10, no. 7, pp. 944 – 954, 2016.
32. G. Mokryani, Y.F. Hu , P. Pillai , H.S. Rajamani, "Active Distribution Networks Planning with High Penetration of Wind Power", Renewable Energy, available online at: http://dx.doi.org/10.1016/j.renene.2016.12.007.
33. Distributed Generation and Sustainable Electrical Energy Centre. United Kingdom Generic Distribution System (UKGDS). [Online]. Available: http://www.sedg.ac.uk

Critical Performance Evaluation of a Few Intelligent Controllers for Effective Power Control in Variable Speed Wind Generators

Rajiv Singh and Asheesh Kumar Singh

Abstract Active and reactive power controls are the two major grid code issues for effective grid integration of wind farms. In this chapter, a novel inverse artificial neural network (ANN) controller has been proposed for effective control of grid side power in a variable speed wind generator. Its performance was tested on a full capacity grid-connected squirrel-cage wind generator. A two-level ac-dc-ac converter was used as buffer between the grid and the wind generator. The inverse ANN controller was designed using neural network system identification (NNSYSID) approach. The performance of inverse ANN controller was compared with adaptive neuro-fuzzy inference system (ANFIS) and a conventional proportional-integral (PI) controller. ANFIS and inverse ANN controllers can cognize the nonlinear dynamics of a plant for performing the control action. A vast amount of input and output data generated from a plant can be used for training while designing these nonlinear controllers. Hence, their design is relatively easier as compared to any other nonlinear controller. A PI controller is designed by linear approximation of a nonlinear plant near an operating point. However, if the operating point of the plant shifts beyond the range of design, the performance of PI controller deteriorates. The use of intelligent ANN and ANFIS controllers can overcome this problem. This study shows the better performance of proposed inverse ANN controller than the ANFIS and PI controllers.

Keywords Neural network controller · PI controller · Variable speed wind generator · Squirrel-cage wind generator

R. Singh (✉)
Department of Electrical Engineering, G.B. Pant University of Agriculture and Technology, Pantnagar, India
e-mail: rajiv77singh@gmail.com

A.K. Singh
Department of Electrical Engineering, MNNIT Allahabad, Allahabad, India

B. Azzopardi (ed.), *Sustainable Development in Energy Systems*,
DOI 10.1007/978-3-319-54808-1_7

1 Introduction

Wind power has in the past two decades emerged as a credible source for generation of electricity. The technological advancements in the wind turbines, power electronic equipment, and controllers have brought the wind generators at par with the conventional generators. Several intelligent and robust control techniques for the control of parameters such as power and frequency, and power factor in grid-connected wind generators have been reported in the past. It is required to control these parameters such that the wind generators comply with the standard grid code conditions of the utility, where they are connected. In spite of several intelligent schemes reported earlier, the quest for new and more reliable controllers has not ended. The work presented in this chapter is an outcome of this quest. Here, a few intelligent inverse ANN and ANFIS controllers are designed for controlling the grid side real and reactive power in a variable speed wind generator. The conventional PI controllers are generally designed by approximating a nonlinear plant with a linear model. The wind generators together with power electronic converters exhibit nonlinear dynamics. So, a controller based on linear model of wind generator may be manifested with severe performance limitations, especially when the range of operation goes beyond the linear range of design. Inverse ANN and ANFIS controllers are relatively easier to design as compared with their counterparts. This is because they can be designed by using the input and output data generated from a plant model.

Before proceeding with the actual design task, an allusion on the pre-existing controllers for control of power in various configurations of variable speed wind generators is justified here. Lin et al. [1] proposed neural and fuzzy controllers for a. c. and d.c. electric drives. Their scheme was similar to the techniques used for maximum power point tracking (MPPT) in the wind generators. Krichen et al. [2] had developed a method aimed to impose an acceptable voltage profile and to reduce active losses of an electrical supply network including wind generators in real time. While, Ko et al. [3] compared the performance of a PI and ANN controller for power control in a DFIG. It was observed that the ANN controllers performed better as compared to the PI controllers. Jerbi et al. [4] have adopted a fuzzy logic controller (FLC) for a control of rotor side converter in a DFIG. Similarly, Singh et al. [5] have designed intelligent ANFIS controller for power electronic converters in wind generators. While, Orlando et al. [6] designed and implemented a fuzzy PI controller for a DFIG and compared its performance with conventional PI controllers.

Martinez-Rojas et al. [7] provided PSO-based optimization for reactive power dispatch in wind farms by taking into account the reactive power at PCC. Amimeur et al. [8] proposed a robust nonlinear sliding mode control (SMC) algorithm for dual stator DFIG. The SMC was designed for full range of wind speed between the cut-in and cutout wind speeds. Tang et al. [9] have proposed an ANN controller for reactive power control of wind farms with DFIG. Here, they investigated the online learning and control approach based on adaptive dynamic programming (ADP) for wind farm control and integration with the grid.

Most of the controllers discussed above have been implemented either on a DFIG- or PMSG-based wind generator. A DFIG generally uses reduced capacity power converters capable of handling power equivalent to one-third of the generator power ratings. This is because only rotor power is handled by the power converters. Low power handling capacity of power converters in DFIG reduces the complicacies of controller design. Other configurations of variable speed wind generators involving SCIG use full capacity power convertors. Hence, the design and implementation of controllers for full capacity wind generators are more complex than DFIG. Therefore, in this chapter, a case of SCIG-based full capacity wind generator is considered for the design and implementation of controllers. The ANFIS and ANN controllers are generally compared with their fuzzy counterparts. For generating the rule base in a fuzzy controller, very precise knowledge of a plant is required. Any gap or discrepancy in acquiring the knowledge of plant behavior may lead to inaccurate fuzzy controller designs. The inverse ANN controllers, however, can easily learn the input–output dynamics of a nonlinear system from a given data set and work in accordance with the training provided, when put in unknown situations. Similarly, in ANFIS controllers, the rule base and membership functions can be generated from the input and output data of the plant using adaptive algorithms. Hence, the design of inverse ANN and ANFIS controllers becomes comparatively easier than any other nonlinear controller. Therefore, these controllers have been chosen for study and analysis. Now-a-days, multi-agent based control technologies have unleashed novel dimensions for the smart integration of renewable energy-based generation with the grid.

This chapter is organized as follows: a brief illustration of wind grid codes is presented in Sect. 2 while in Sect. 3, the mathematical models of system components are described. The proposed control scheme and design of controllers are described in Sect. 4. Finally, the results and conclusions are presented in Sects. 5 and 6, respectively. The Appendix contains the necessary data used for simulations, and references are presented in end of the chapter.

2 Wind Grid Code

Grid codes are the technical requirements for connecting and disconnecting the generation and load in a power system [10–12]. These technical requirements and associated technical supports should be strictly followed by the wind farm operators, in order to ensure reliable and qualitative power generation, from a wind power plant. Generally, these technical requirements are the commonly agreed issues decided by the transmission system operators (TSO) and the power plant operators. The objectives underlying these conditions are based on the mutual benefits of both the TSO and power plant operators, without compromising the overall security of the power system. Most of the countries have formulated the wind grid codes according to their specific needs, but essentially their basic framework is similar comprising issues such as (a) low voltage ride through

(LVRT), (b) voltage and reactive power control, and (c) frequency–active power control [13–15].

Active and reactive power controls are the two prominent grid code issues in a grid-connected wind farm. Power control is generally required during (a) wind speed changes, (b) voltage variations, and (c) frequency changes. Active power control is generally defined in terms of ramp rates for the above first two conditions and in the form of frequency-active power curve as depicted in Fig. 1 for the third case. Active power and frequency are related quantities in power system; hence, it is required to regulate the active power according to the changes in system frequency in order to maintain its stability. The allowable ramp rates for active power exhibit worldwide variations, for instance, German grid code allows a ramp rate of 10% per minute of the generator power rating, while the Irish grid code allows a ramp of 1–30 MW/min. But, the Denmark grid code allows a ramp rate of 10–100% of rated power per minute [16].

Reactive power control at a node in a power system is generally required for maintaining the voltage stability. Very small deviation in the node voltage is desired for achieving better voltage stability [17]. Hence, for the control of voltage at a node, it is desired that the generating units supply controlled leading/lagging reactive power at grid connection point. Most of the wind grid codes desire that the reactive power control by wind farms should be accomplished in a manner similar to the conventional generators. Generally, the wind grid codes represent the reactive power requirements in the form of a voltage–power factor curve depicted in Fig. 2.

For appropriate accomplishment of wind grid code requirements, it is desired to use efficient and robust controllers for active and reactive power control. More accurate and appropriate controllers adopting robust algorithms for the control of active and reactive power are required in view of the increasing stiffness in the grid codes. Here, a few intelligent techniques for power control have been analyzed and compared.

Fig. 1 Frequency–active power characteristics of wind generators

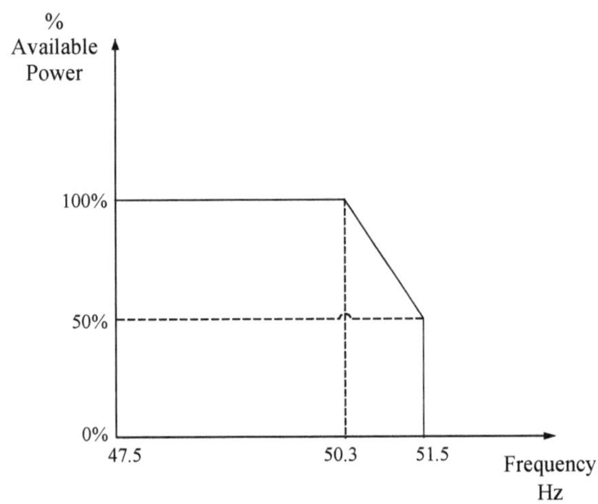

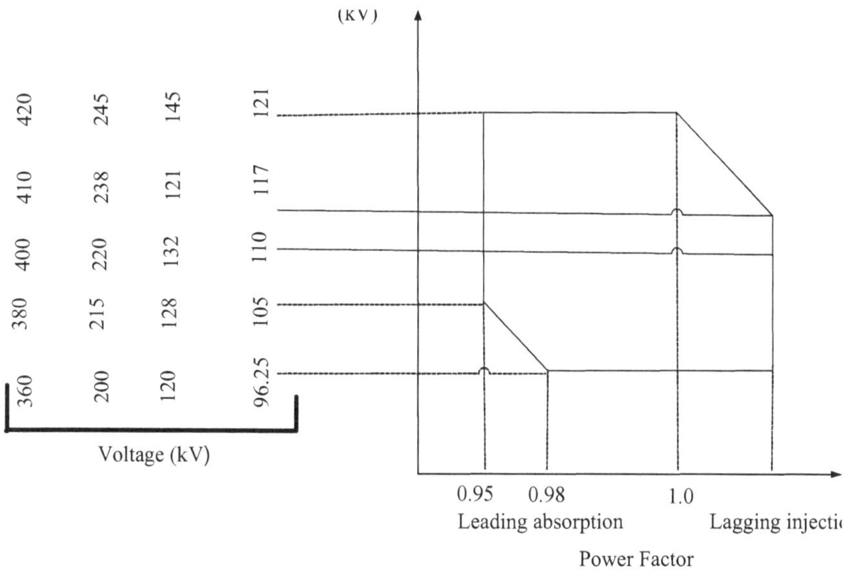

Fig. 2 Voltage–power factor characteristics of wind generators

3 System Description and Modeling

Figure 3 depicts a simplified diagram of the test system considered in this chapter. The wind generator contains a three-blade wind turbine coupled with a squirrel-cage induction generator (SCIG) through a gearbox. The SCIG power, voltage, and frequency ratings are 30 kW, 460 V (L–L), and 60 Hz, respectively. The turbine coupled with the generator operates at optimum pitch angle and delivers rated power at base wind speed of 12 m/s. The equivalent circuit parameters of SCIG are given in the appendix. An appropriate value of capacitor bank not included in Fig. 3, is connected at the SCIG stator terminals for providing reactive power support. The output power from generator is fed to the grid through two VSCs connected by a d.c. link capacitor and line impedances. The VSC connected toward the generator is called generator side converter and the one connected toward grid is called grid side converter (GSC). These VSCs are capable of handling rated power produced by the generator. Hence, this kind of wind generator configuration is also sometimes called full capacity wind generator. The first VSC is used for speed control of the generator, and GSC is used for power control at PCC. The purpose of speed control is for MPPT by the wind generator. Vector control scheme is used here for MPPT. The line impedance represents resistance and reactance of power cables, and other equipment such as transformers connected

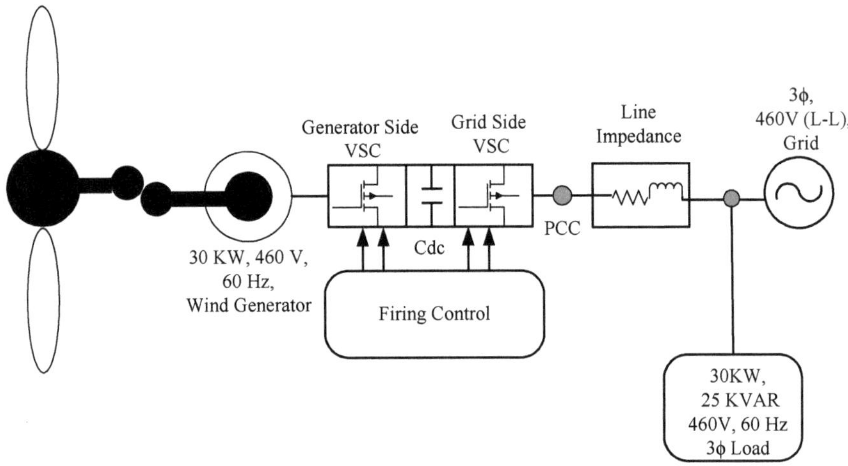

Fig. 3 Grid-connected variable speed wind generator system

between the generator and the grid. The grid operates at a balanced 3-ϕ voltage of 460 V (L–L) r.m.s. and a frequency of 60 Hz. The grid is a weak grid with short circuit capacity of 100 MVA and X/R ratio 10. A three-phase load of 30 kW and 25 KVAR is also connected at PCC. The mathematical models of wind turbine, induction generator, and the VSCs are described in the subsequent sections.

3.1 Wind Turbine Model

The aerodynamic power output from wind turbine is modeled using (1)

$$P_{\text{mech}} = 0.5\rho A_{\text{wtr}} V_{\text{w}}^{3} C_{\text{p}}(\lambda, \beta) \tag{1}$$

where P_{mech} is the mechanical power developed by the wind turbine rotor, ρ is the air density in kg/m^3, A_{wtr} is the area swept by the wind turbine rotor in m^2, V_{w} is the velocity of blowing wind in m/s, $C_{\text{p}}(\lambda, \beta)$ is the power coefficient also called Betz coefficient.

P_{mech} depends mainly on C_{p}, which is considered as a very important design parameter affecting the efficiency of a wind turbine. The maximum theoretical value of C_{p} is 0.59. Its value depends on λ (tip-speed ratio) and β (blade pitch angle). Here, a fixed pitch turbine is considered with blade pitch angle $\beta = 0°$. From the turbine characteristics depicted in Fig. 4, it can be observed that for every wind speed the power output from turbine is maximum at a particular turbine speed [18]. The expression for $C_{\text{p}}(\lambda, \beta)$ is given by (2).

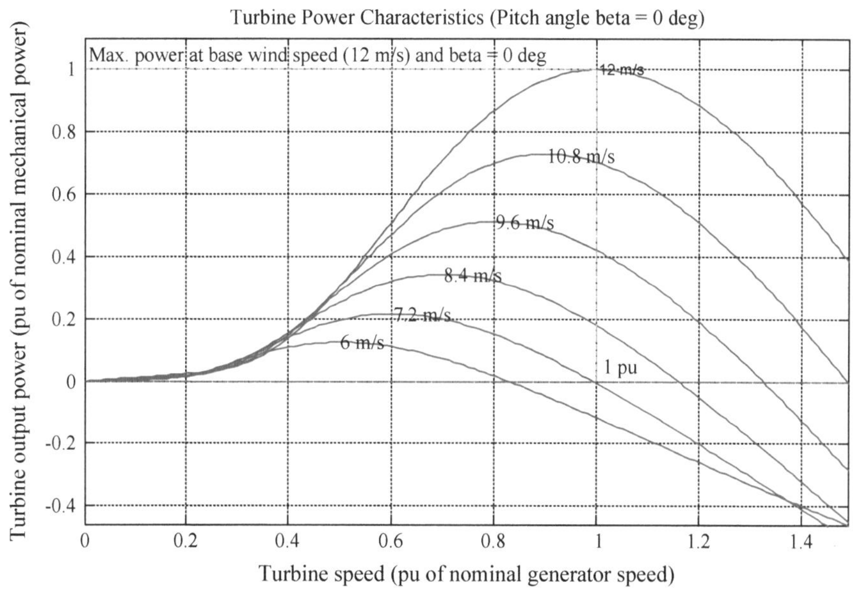

Fig. 4 Wind turbine power characteristics

$$C_p(\lambda, \beta) = C_1\left(\frac{C_2}{\lambda_i} - C_3\beta - C_4\right)e^{\frac{-C_6}{\lambda_i}} + C_8\lambda \tag{2}$$

where the values of constants $C_1 - C_6$ are $C_1 = 0.5176$, $C_2 = 116$, $C_3 = 0.4$, $C_4 = 5$, $C_5 = 21$, and $C_6 = 0.0068$, respectively, and the intermediate term λ_i is given by (3).

$$\frac{1}{\lambda_i} = \frac{1}{\lambda + 0.08\beta} - \frac{0.035}{\beta^2 + 1}. \tag{3}$$

3.2 *Drive Train Model*

In a wind turbine, there are several rotating components such as blades, the turbine hub, gears, and the rotor of coupled generator. Some of these components rotate at higher speeds, while others rotate at lower speeds due to gearbox. These components are generally termed as drive train. As these components rotate, the associated mechanical quantities such as inertia, spring constants, and damping coefficients interact with one another, thereby affecting the power output from the turbine. The drive train dynamics incorporate low frequency oscillations in the output torque of the wind turbine. An oscillatory torque consequently leads to oscillations in the electric power output from the generator. The drive train model is generally connected between the turbine aerodynamic model and the generator model [19]. There are several drive train models such as one-mass model, two-mass model, three-mass model, and six-mass model used in previous works. Figure 5 depicts the most accurate six-mass drive train model used in this work and the symbols for the model are explained in Table 1. In a six-mass model, the drive train contains six rotating masses represented by their respective inertia. The inter-component torsions are represented by spring constants, and both the self and the inter-component damping are represented by damping coefficients. All the quantities in the model such as inertia, spring constants, and damping coefficients are referred to uniform speed of rotation using (4).

$$\frac{I_{\text{hs}}}{I_{\text{ls}}} = \frac{S_{\text{hs}}}{S_{\text{ls}}} = \frac{\tau_{\text{hs}}}{\tau_{\text{ls}}} = \frac{N_{\text{hs}}^2}{N_{\text{ls}}^2} \tag{4}$$

where I_{hs} is the inertia on high speed side, I_{ls} is the inertia on low speed side, S_{hs} is spring constant on high speed side, S_{ls} is the spring constant on low speed side, τ_{hs} is the damping on high speed side, τ_{ls} is the damping on low speed side, N_{hs} and N_{ls} are the speeds on the high and low speeds sides, respectively.

The above drive train model was simulated in Simscape, which is the physical system modeling tool in Simulink. In this model, the total friction has been considered to be referred on the generator side. It is also assumed that each of the three turbine blades share equal aerodynamic torque. Effect of unequal torque sharing by the blades can also be explored; however, it has not been included in this work.

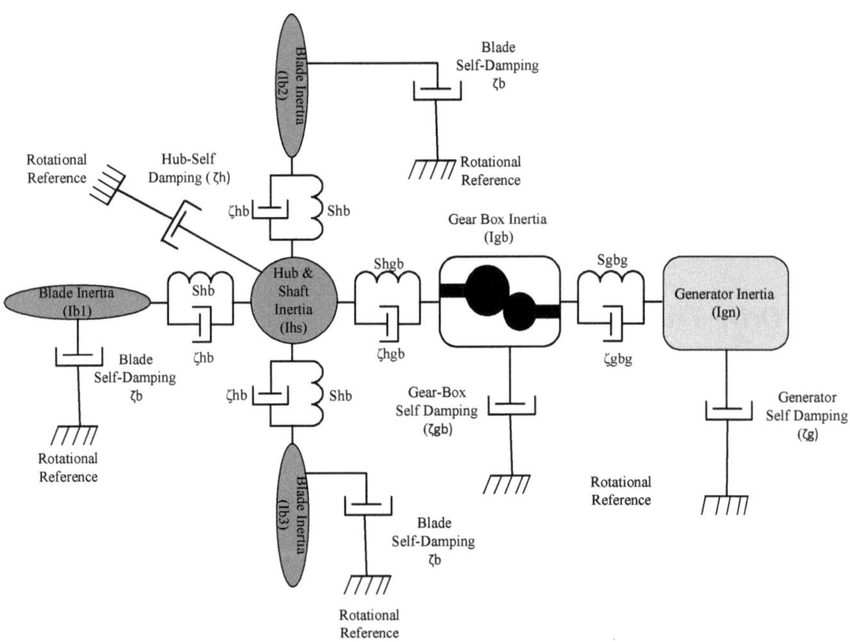

Fig. 5 Six-mass drive train model [19]

Table 1 Symbol description for drive train model

S.No.	Symbol	Description
1.	ζ_{ik} and S_{ik}	Mutual damping and spring constant between *i*th and *k*th inertia
2.	ζ_m and S_m	Self damping and spring constant of *m*th inertia
3	I	Inertia
4.	Subscripts b, h, gb, and gn represents blade, hub, gearbox and generator, respectively.	

3.3 Generator Model

The generator used in this work is a SCIG modeled in *d–q* (synchronous) frame of reference. For obtaining the equations, it is assumed that the stator and the rotor windings are sinusoidal and symmetrical [20]. The model is described mainly by three set of equations (a) voltage equations, (b) flux linkage equations, and (c) equation for motion. The *d–q* axis voltage equations used for modeling are given in (5).

$$\left.\begin{array}{ll} V_{ds} = R_s I_{ds} + D\phi_{ds} - \omega\phi_{qs}; & V_{qs} = R_s I_{qs} + D\phi_{qs} + \omega\phi_{ds} \\ V_{dr} = R_r I_{dr} + D\phi_{dr} - (\omega - \omega r)\phi_{qr}; & V_{qr} = R_r I_{qr} + D\phi_{qr} + (\omega - \omega r)\phi_{dr} \end{array}\right\} \quad (5)$$

The flux linkage equations are given in (6)

$$\left.\begin{array}{ll} \phi_{ds} = L_s I_{ds} + L_m I_{dr}; & \phi_{qs} = L_s I_{qs} + L_m I_{qr} \\ \phi_{dr} = L_r I_{dr} + L_m I_{ds}; & \phi_{qr} = L_r I_{qr} + L_m I_{qs} \end{array}\right\} \quad (6)$$

The electromagnetic torque equation and the equation for motion are given by (7) and (8), respectively [20].

$$\left. T_m = \frac{3PLm}{2Lr}\left(I_{qs}\phi_{dr} - I_{ds}\phi_{qr}\right)\right\} \quad (7)$$

$$\left. JD\omega_m = T_e - T_m\right\} \quad (8)$$

where V_{ds}, V_{qs}, V_{dr}, V_{qr} are *d–q* axis stator and rotor voltages; I_{ds}, I_{qs}, I_{dr}, I_{qr} are *d–q* axis stator and rotor currents; ϕ_{ds}, ϕ_{qs}, ϕ_{dr}, ϕ_{qr} are *d–q* axis stator and rotor fluxes; R_s, R_r are stator and rotor winding resistances; ω, ω_r, ω_m are rotating speed of arbitrary reference frame, rotor electrical angular speed and rotor mechanical speed, respectively; D is the differentiation d/dt; L_s, L_r, L_m are stator self-inductance, rotor self-inductance and magnetizing inductance; J is moment of Inertia; P is number of pole pairs; and T_m, T_e are mechanical torque from generator shaft and electromagnetic torque.

3.4 Power Converter Model

A two-level converter depicted in Fig. 6 is an ac-dc-ac converter, which contains six unidirectional commanded IGBTs functioning as rectifier and equal number of unidirectional commanded IGBTs as an inverter [21]. Hence, there are six IGBT switches for the rectifier as well as for the inverter. In each switch, a freewheeling diode is also connected antiparallel with the IGBT. The switches are arranged on the legs of the converter. So both the rectifier and the inverter have three legs with two switches connected on each leg. The legs also represent three phases of the power converter. The IGBTs connected on a leg k of the converter are identified by index $i = 1, 2$. Hence, each IGBT switch on a leg can be denoted by a pair of indexes (i, k). A function S_{ik} (switching function or conduction state) represents the state of conduction for the IGBT switches. A switching variable γ_{k} is used to activate the switching function S_{ik} for deciding the conduction state of the IGBT i on leg k of the converter. Respectively, the index k with $k \in \{1, 2, 3\}$ denotes the rectifier legs and $k \in \{4, 5, 6\}$ identifies legs of the inverter. The switching variable for a leg is the function of logical conduction states (switching function) given by (9) [21].

$$\gamma_k = \left\{ \begin{matrix} 1, & (S_{1k} = 1 & \& & S_{2k} = 0) \\ 0, & (S_{1k} = 0 & \& & S_{2k} = 1) \end{matrix} \right\} k \in \{1, \ldots, 6\} \tag{9}$$

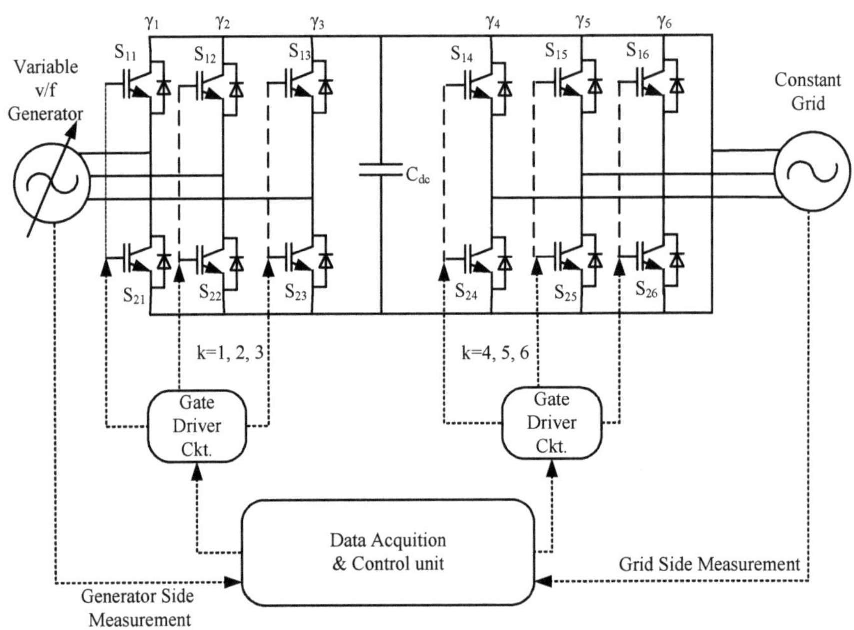

Fig. 6 Two-level converter

However, the logical conduction states are given by (10).

$$\sum_{i=1}^{2} S_{ik} = 1 \quad k \in \{1, \ldots, 0.6\} \tag{10}$$

The logical conduction states (S_{ik}) depend on the conduction and the blocking states of the IGBTs.

The voltage of capacitor (V_{dc}) is modeled by (11).

$$\frac{dv_{dc}}{dt} = \frac{1}{c}\left(\sum_{k=1}^{3} \gamma_k i_k - \sum_{k=4}^{6} \gamma_k i_k\right) \tag{11}$$

Hence, a two-level converter can be modeled by (9)–(11).

4 Control Scheme and Design of Controllers

For a two-level GSC used in the wind power system represented in Fig. 3, the current control scheme as depicted in Fig. 7 is adopted. The power being injected into the grid by the wind generator is controlled by regulating the 3-ϕ GSC currents. The objective here is to control the active and reactive power at the grid connection point (PCC) in the wind generator system. The decoupled *P*–*Q* control strategy is implemented in α–β reference frame. Using the reference power commands ($P_{ref.}$ & $Q_{ref.}$), the current reference signals in α–β reference frame are generated according to (12) and (13), respectively [22]. The currents I_α and I_β at the GSC terminals are measured and compared with the reference currents $I_{\alpha ref.}$ and $I_{\beta ref.}$ respectively. The corresponding errors after comparison are fed to a controller, which can either be a conventional PI controller or their intelligent counterparts such as the inverse ANN and ANFIS controller. The controller outputs are transformed from α-β reference frame to a–b–c frame and given to a pulse width modulator (PWM). It produces modulating signals for deciding the firing sequence of the IGBT switches. The firing sequence is controlled in such a way that the values of output currents are regulated satisfactorily.

$$I_{\alpha ref.} = \frac{2}{3(V_{\alpha pcc}^2 + V_{\beta pcc}^2)}(V_{\alpha pcc}P_{ref} + V_{\beta pcc}Q_{ref}) \tag{12}$$

$$I_{\beta ref.} = \frac{2}{3(V_{\alpha pcc}^2 + V_{\beta pcc}^2)}(V_{\alpha pcc}P_{ref} - V_{\beta pcc}Q_{ref}) \tag{13}$$

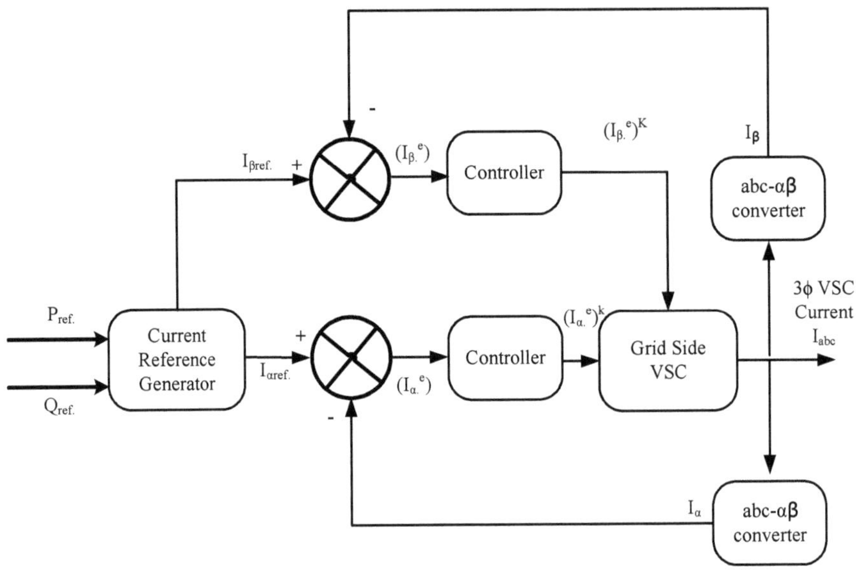

Fig. 7 Current control scheme for two-level VSC

4.1 Inverse ANN Controller

Evolutions in computational intelligence (CI) have produced novel approaches for design of intelligent controllers, especially where the conventional methods fail in realizing the appropriate plant/process dynamics for computing an appropriate control law [23]. ANN offers alternative approach for synthesizing effective non-linear controllers by precisely learning the nonlinear plant dynamics from the information obtained using experiments on a plant [23]. In this work, a novel direct inverse ANN control method with low computational complexity is introduced. This method involves training of ANN to mimic the inverse of a plant dynamics. As the plant and its inverse dynamics connected in cascade cancel out, the resultant unity transfer function causes the plant to follow the reference input. Here, a single ANN model is used as control law which calculates the current values of manipulated variables fed as input to the plant at discrete time intervals. The past values of input and output variables in a plant are used for training the ANN model, and desired values are used as reference.

4.1.1 Inverse Dynamic Model for Controller Design

For a SISO system described by 'n' state variables, a discrete dynamic model correlating the next value of output (controlled) variable with the current values of state variable and the input (manipulated variable) is given by (14).

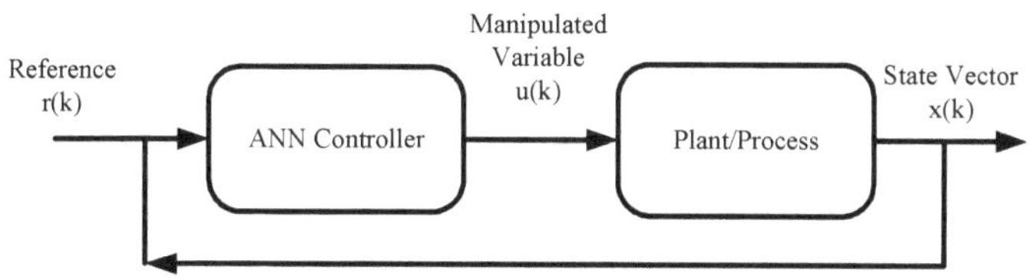

Fig. 8 A neural controller based on inverse plant model

$$y(k+1) = f(x(k), u(k)) \tag{14}$$

where y is the output, x is the state vector, u is the input, and f is a nonlinear function. The system in (14) can be described by inverse dynamics given in (15).

$$u(k) = \psi(x(k), y(k+1)) \tag{15}$$

where ψ is the function representing the inverse dynamics of the plant in (14). In (15), if we replace the next value of controlled variable $y(k + 1)$ with the current value of reference $r(k)$, a control law in (16) can be obtained, which can decide the value of manipulated variable to be applied as input at any discrete time instant, so that process is driven from its current state to the reference value.

$$u(k) = \psi(x(k), r(k)) \tag{16}$$

The unknown function ψ (inverse dynamics) can be approximated with the help of a neural network, which can be used as a controller. The controller receives the values of reference and state variables at each discrete time instant for calculating the value of manipulated variable to be applied to the system at the next instant [24]. The operation of inverse ANN controller is depicted in Fig. 8.

4.1.2 System Identification Approach and Nonlinear ARX Model Estimation

The process of constructing models of a dynamic system from the experimental data is called *system identification*. This approach for obtaining the system model is used for complicated systems, where the first principles of physics for creating models become cumbersome. The information about system dynamics is obtained from the experiments conducted on a plant using measured outputs by applying known inputs. System identification generally involves three steps (a) experimental planning, (b) model structure selection, and (c) parameter estimation for obtaining a plant model. Different approaches for system identification are used with the help of different data forms obtained from the system. When only input–output observations are used, behavioral or black box model can be constructed. In black box modeling, neural networks play an important role. A number of neural network techniques such as back-propagation network, Hopfield network, and Kohonen network have been used in system identification. The back-propagation neural networks applied here can be used to empirically map any function using measured experimental data. The estimated models depicted in Fig. 12 are nonlinear ARX models obtained by extending the linear ARX model structures given by (17) and (18).

$$\begin{aligned} y(t) &= a_1 y(t-1) + a_2 y(t-2) + \cdots + a_{\mathrm{na}} y(t-na) \\ &= b_1 u(t) + b_2 u(t-1) + \cdots + b_{\mathrm{nb}} u(t-nb+1) + e(t) \end{aligned} \tag{17}$$

$$\begin{aligned} &y_p(t) = [-a_1, -a_2, \ldots, -a_{\mathrm{na}}, b_1, b_2, \ldots b_{\mathrm{nb}}] * \\ &[y(t-1), y(t-2), \ldots, y(t-na), u(t), u(t-1), \ldots, u(t-nb-1)]^T \end{aligned} \tag{18}$$

where $y(t-1)$, $y(t-2)$, … , $y(t-\mathrm{na})$, $u(t)$, $u(t-1)$,…,$u(t-nb-1)$ are delayed input and output variables called regressors, and a_1, a_2,…, a_{na}, b_1, b_2,…, b_{nb} are weighted coefficients of outputs and inputs, respectively. From the above equations, it can be observed that the current output $y(t)$ is calculated from the past input and output values and the current input values. The above model can be extended to create a nonlinear ARX model in (19).

$$y_p(t) = f(y(t-1), y(t-2), y(t-3), \ldots, u(t), u(t-1), u(t-2), \ldots) \tag{19}$$

where f is nonlinear mapping function having model regressors as the input.

A nonlinear ARX model can be generated in two stages as depicted in Fig. 9. The first step involves computation of regressors from the current and past input values along with the past output data. The second stage which is nonlinearity estimator contains a linear function in parallel with a nonlinear function. The nonlinearity estimator block maps regressors to the model output using a combination of both these functions. If the nonlinearity estimator is described by (19)

$$f(x) = L^{\mathrm{T}}(x-r) + d + g(Q(x-r)) \tag{20}$$

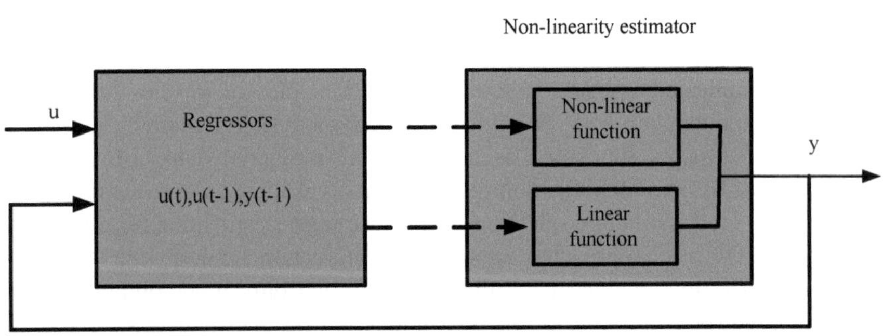

Fig. 9 Nonlinear ARX model in simulation environment

where x is a vector of the regressors, $L^{T}(x - r) + d$ is the output of the linear function block and d is the scalar offset, $g(Q(x - r))$ is the output of nonlinear block, r is the mean of the regressors, x and Q are the projection matrix [24, 25]. While estimating the nonlinear ARX model, its parameter values such as L, r, d, and Q are computed using the pre-specified parameter g. In this work, ANN described in the next section is used as the nonlinearity estimator.

4.1.3 Controller Design Using NNSYSID Approach

Here, NNSYSID approach is adopted for the design of inverse ANN controller. A feed-forward ANN with one hidden layer containing 10 neurons and one input/output layer each containing single neuron was created for this purpose. The training data for system identification was obtained from the simulated plant model depicted in Fig. 3 on the basis of implemented control scheme presented in Fig. 7 using PI controllers. The same input–output data set was used for designing the controllers for both α and β current control loops as shown in Fig. 7. Randomly varying signals representing errors (I_{α}^{e}) and (I_{β}^{e}) in α and β components of GSC currents were given as input to the plant model and the corresponding GSC terminal currents I_{α} and I_{β} were taken as output signal. In order to capture the plant dynamics effectively, the data in fixed samples was obtained for a wide range of operation. The input and output data thus obtained was used to identify the inverse ARX model of the system using ANN with the help of system identification toolbox in MATLAB according to (17)–(20). Preprocessing techniques such as detrenting, which involves removing means and linear trends in the data, were applied to bring the data in its final shape. For training of the ANN using back-propagation technique, the output data from the plant was used as input to the ANN and input data to the plant was taken as output. Then, the trained ANN was called into the Simulink model for testing its performance as controller. The input–output data for training and testing are depicted in Figs. 10 and 11, respectively, while Fig. 12 shows the various ARX inverse system models identified using ANN.

From Fig. 12, it can be observed that first curve (Test Data) represents the reference data for testing and validating the identified models. The second curve (Model 1) represents the first identified system representing the inverse plant dynamics. The accuracy of Model 1, when compared with test data, is just 60.91%, and hence, it is not suitable to be used as a controller. The curve Model 2 exhibits 85.36% accuracy. This model was obtained by considering Model 1 as initial model. Similarly, Model 3 obtained by considering model 2 as initial model is 87.16% accurate. The similar identification procedure was followed for obtaining Model 4 and Model 5 with the accuracies of 95.03 and 86.7%, respectively.

Model 4 being the most accurate, is most suitable to be used as controller in this study. The performance of Model 4 is tested by calling the trained *ANN* object into the Simulink plant model from the MATLAB workspace. The test results are presented in Sect. 5.

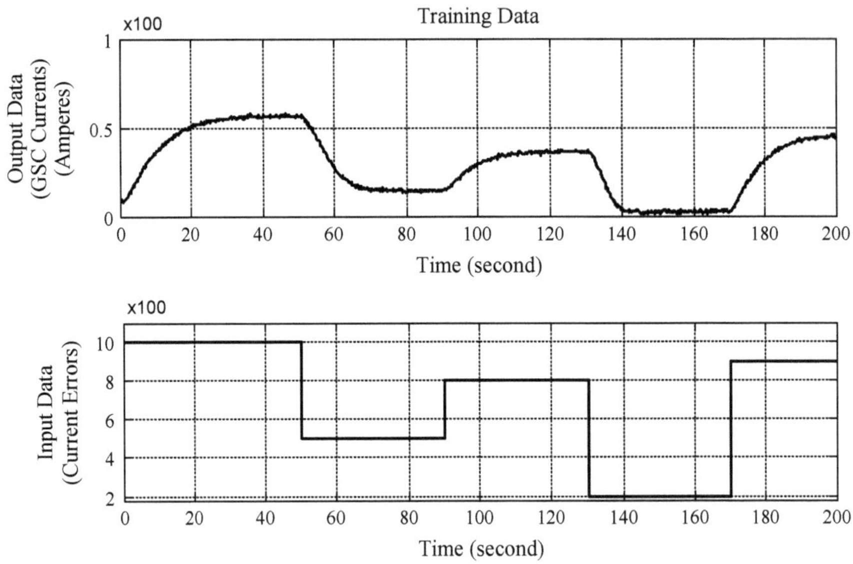

Fig. 10 Training data for inverse ANN controller

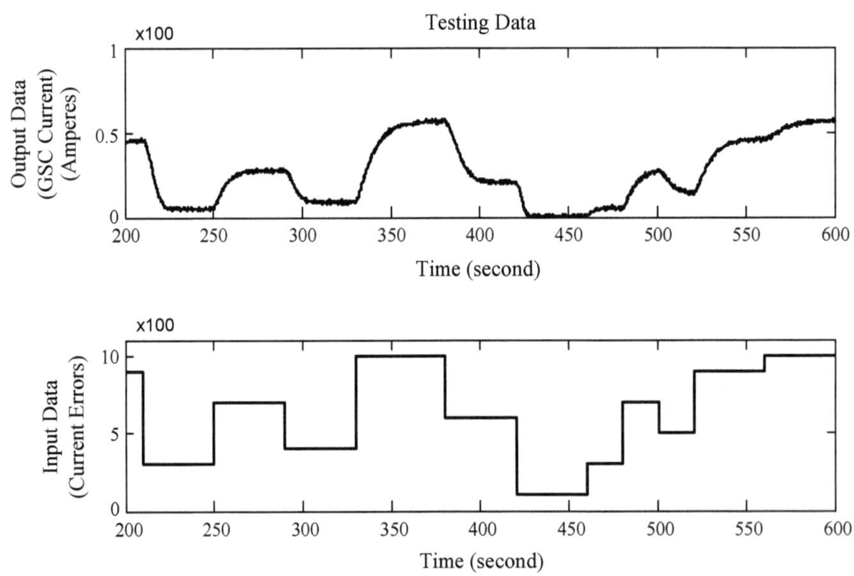

Fig. 11 Testing data for inverse ANN controller

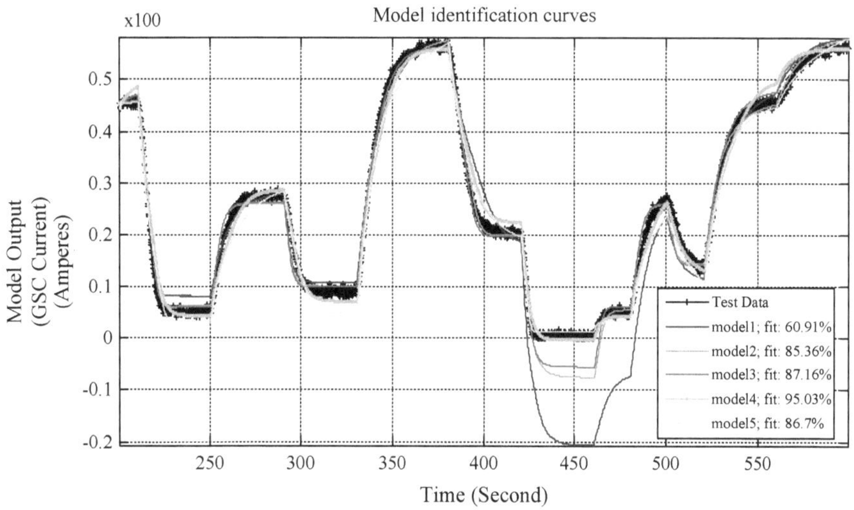

Fig. 12 Inverse system models obtained using the ANN

4.2 Adaptive Neuro-Fuzzy Inference (ANFIS) System

It is a hybrid technique involving ANN and fuzzy logic for creating a fuzzy inference system (FIS). The neural networks possess excellent learning capability from a set of input–output data and perform accordingly on the basis of gained knowledge. The process of learning is alleged to be similar to our brain [24]. But, the information processing in our brain is quite different from ANN. A trained ANN itself can be used as controller for regulating a desired quantity in a plant. However, the ANN is more or less representative of a black box type system cognizing the information as weights and biases relatively difficult to understand. The fuzzy logic is more explanatory and easy to understand than ANN. This is because the representation of information is in the form of linguistic (if-then) rules. The fuzzy systems, however, are manifested with a disadvantage that they cannot learn like ANN. In order to explore the 'self-learning' character of ANN and 'easy representation' character of fuzzy systems, a hybrid ANFIS scheme is generally used. The ANFIS systems can substitute ANN and fuzzy systems in almost every application; hence, they can also be used to control plants or processes [26, 27].

4.2.1 Description and Design of ANFIS Controller

The ANFIS controller used for controlling active and reactive power in a grid-connected wind generator system is depicted in Fig. 13. The controller supports Sugeno-type FIS containing four major components, i.e., a fuzzifier, data set

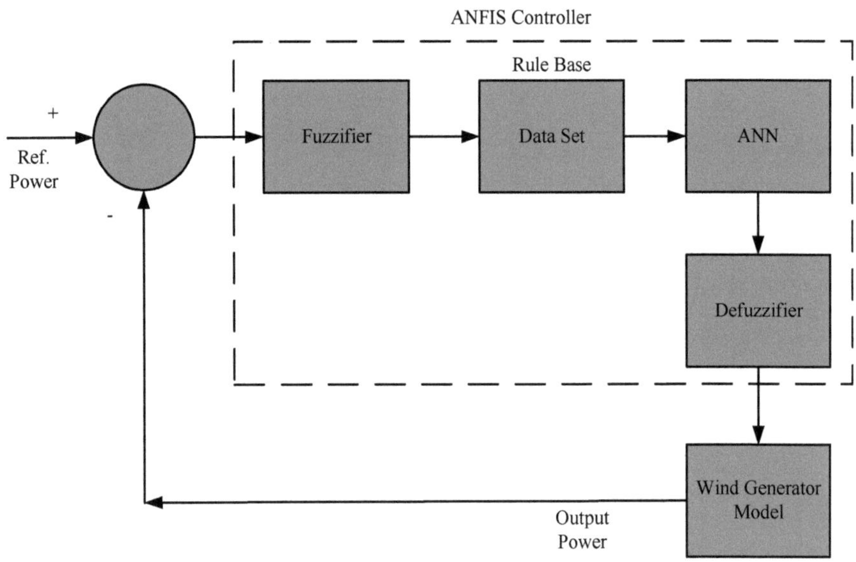

Fig. 13 Block diagram of ANFIS control for variable speed wind generator

as knowledge base, the ANN for training and finally a defuzzifier. The controller accepts two inputs, i.e., error and change in the error are given by (21) and (22).

$$e(n) = I_{\alpha}(\text{ref}) - I_{\alpha}(t) \quad (21)$$

$$\Delta e(n) = e(n) - e(n-1) \quad (22)$$

The fuzzifier unit converts crisp data into a set of linguistic rules. The defuzzifier block is responsible for conversion of non-crisp information (rule base) into numerical (crisp) output. The design procedure of ANFIS controller is depicted in Fig. 14. The training data was obtained from the conventional PI controller by the transient simulation of wind generator system depicted in Fig. 3. The previously obtained input/output data as depicted in Figs. 10 and 11 along with the data for change in error (Δe) generated from simulation was used for ANFIS design. Like the input/output data, the change in error data was also preprocessed for removing the anomalies in it. Similar to inverse ANN design case, the data for change in error was also generated in a very wide operating range, so as to capture even the minute dynamics of the plant.

The initial FIS with 49 rule base was generated from the available data using sub-clustering approach. Triangular membership functions depicted in Figs. 15a, b were chosen for the inputs. An initial rule base was generated from the data set and used by the ANN block for training. ANN was trained using the available data set to modify the initial rule base such that the input and output conditions were satisfied. Here, again back-propagation technique was used for ANN training. The

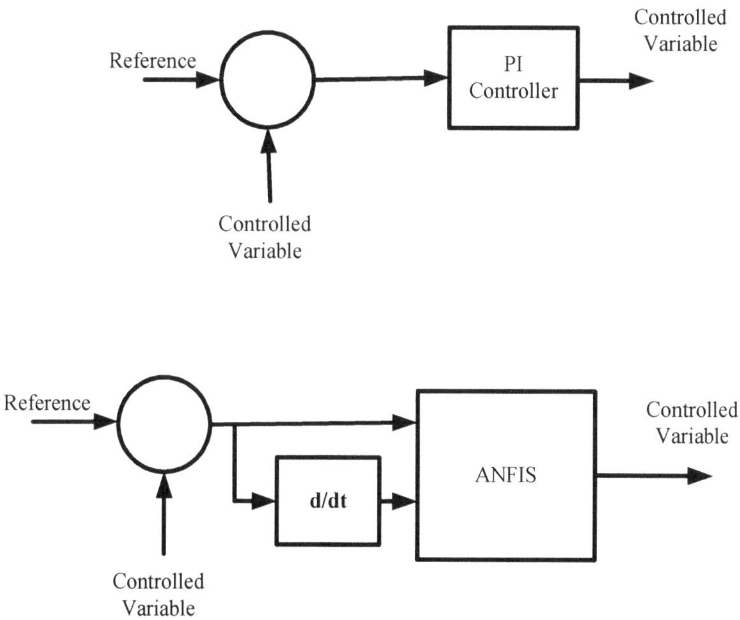

Fig. 14 Design of ANFIS controller

tuning of initial rule base during training stage is very important part of ANFIS design for appropriate rule base selection. The defuzzifier block is used for conversion of non-crisp information (rule base) into numerical (crisp) output. The surface generated for the initial FIS is shown in Fig. 16a, and the surface generated after ANFIS training is depicted in Fig. 16b.

5 Results and Discussion

The results obtained from simulation are presented in this section. The performance of designed controllers has been tested for a step change in the wind speed as depicted in Fig. 17. Step function exhibits a sudden increase or decrease in the value of a physical quantity and, hence, is most suitable for testing the performance of controllers. The wind speed suddenly drops from base speed (12) to 10.8 m/s at 0.1 s. Consequently the value of mechanical torque produced by the wind turbine also decreases from base value (−195 N.m) to −137 N.m, as depicted in Fig. 18. The negative sign indicates the working of machine model in generating mode. It is required to control the turbine speed for MPPT at the reduced value of wind speeds. Moreover, for grid code accomplishments, the active power at PCC should also be changed as quickly as possible corresponding to the change in wind speed. The turbine reference speed and reference active power signals corresponding to a

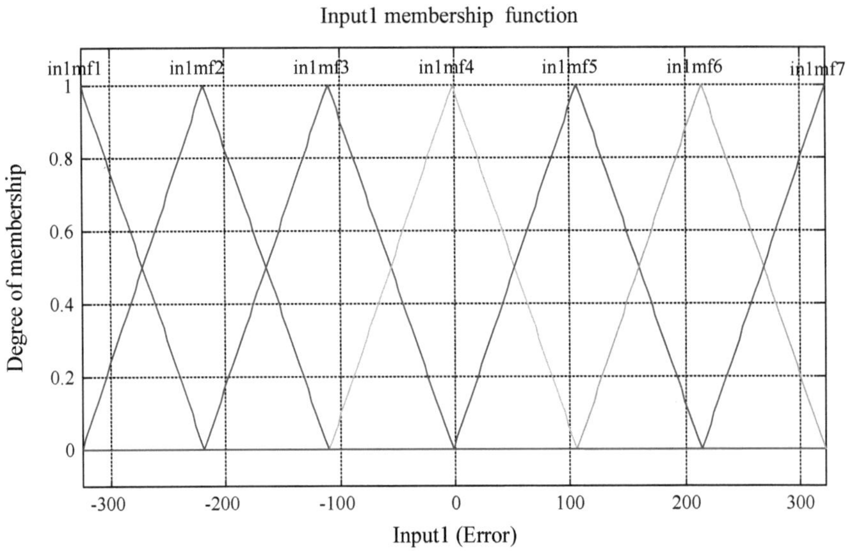

(a) Membership function for input1.

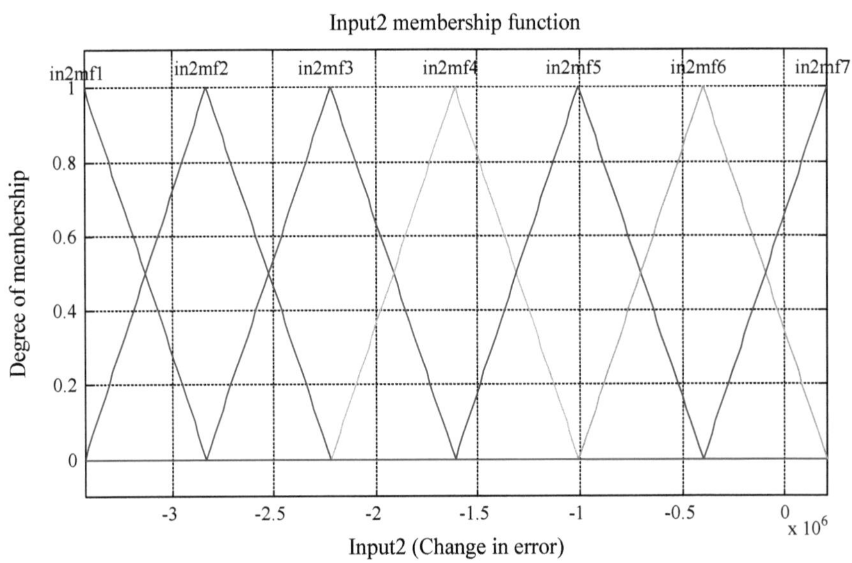

(b) Membership function for input2.

Fig. 15 **a** Membership function for input1 **b** membership function for input2

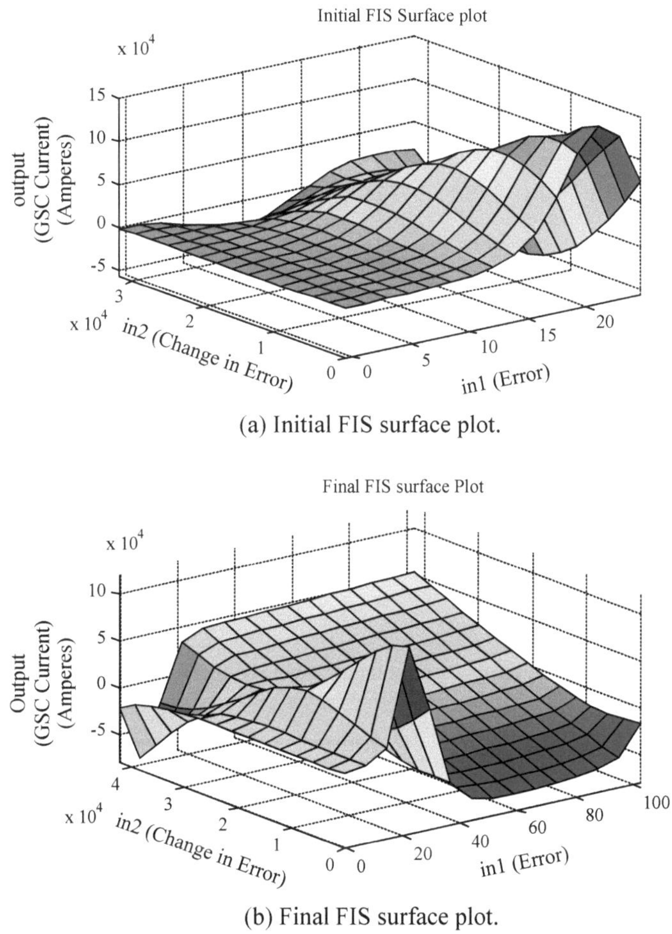

Fig. 16 **a** Initial FIS surface plot **b** final FIS surface plot

particular value of wind speed are obtained from the turbine power curves shown in Fig. 4. These values are represented in Table 2.

The quantities such as turbine rotor speed and machine electromagnetic torque are depicted in Figs. 19 and 20, respectively. As the wind speed decreases from 12 to 10.8 m/s., it is desired that the turbine speed also changes from 193 to 173.7 rad/s. as quickly as possible for MPPT. It can be observed from Fig. 19; the change is achieved at 0.275 s. Unlike the reference (step) signal, the change in speed is gradual (ramp function). This is due to the inertia constant of the turbine rotor. The drop in the rotor speed is due to the corresponding increase in the electromagnetic torque from 195 to 300 N.m at 0.1 s. Figures 21 and 22 depict the controller performance designed in the previous sections. As the wind speed changes from 12 to 10.8 m/s, it is desired that the value of active power at PCC also gets reduced from 30 kW (rated value) to 21 kW. From Fig. 21, it can be observed

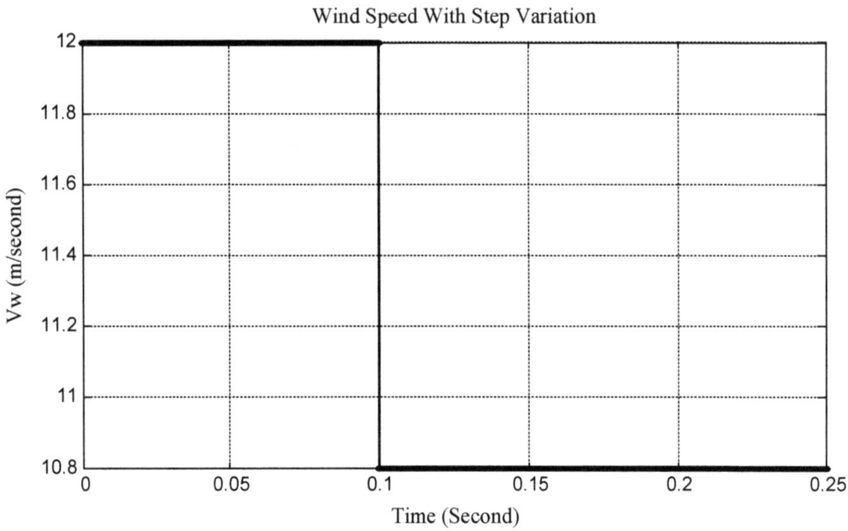

Fig. 17 Varying wind speed characteristics with step variation

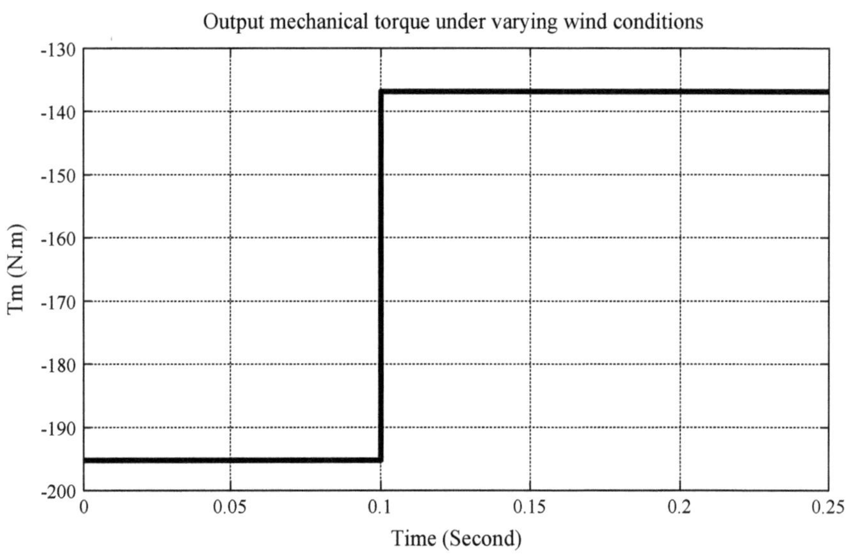

Fig. 18 Output mechanical torque with varying wind conditions

Table 2 Values of reference turbine speed and active power at different wind speeds

S. No.	Wind speed (m/s)	Ref. rotor speed (rad/s)	Ref. active power (kW)
1	12	193	30
2	10.8	173.7	21

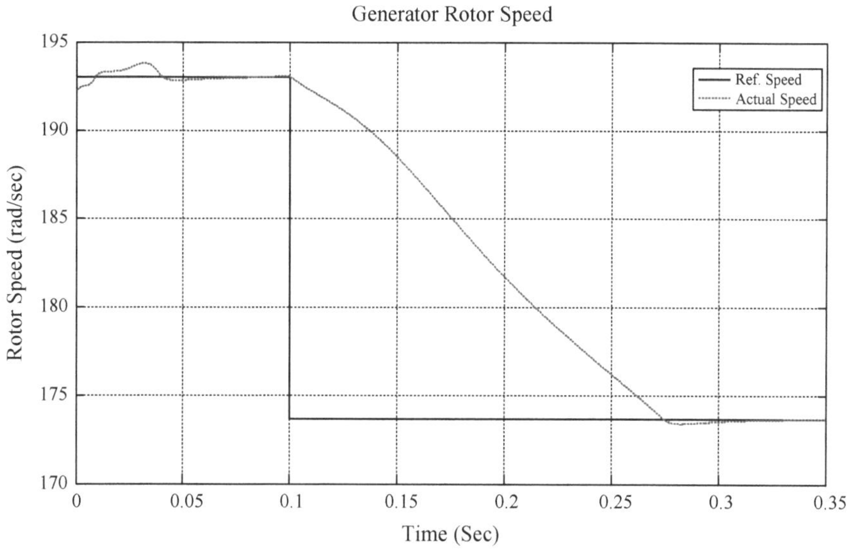

Fig. 19 Turbine rotor speed

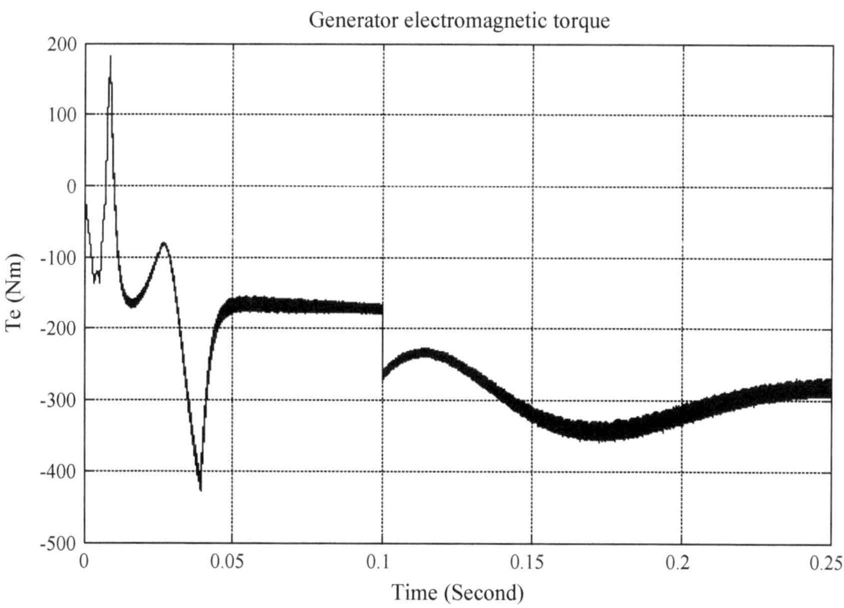

Fig. 20 Generator electromagnetic torque

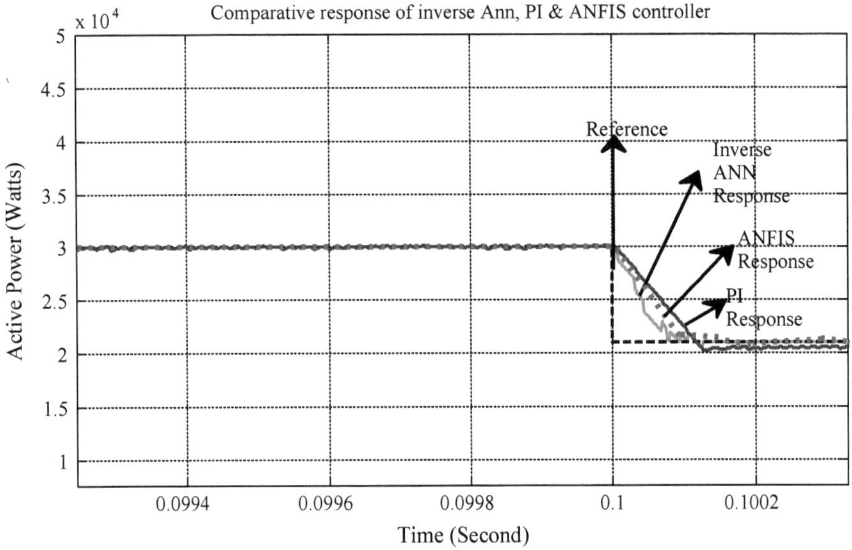

Fig. 21 Comparison of controller performance for active power control

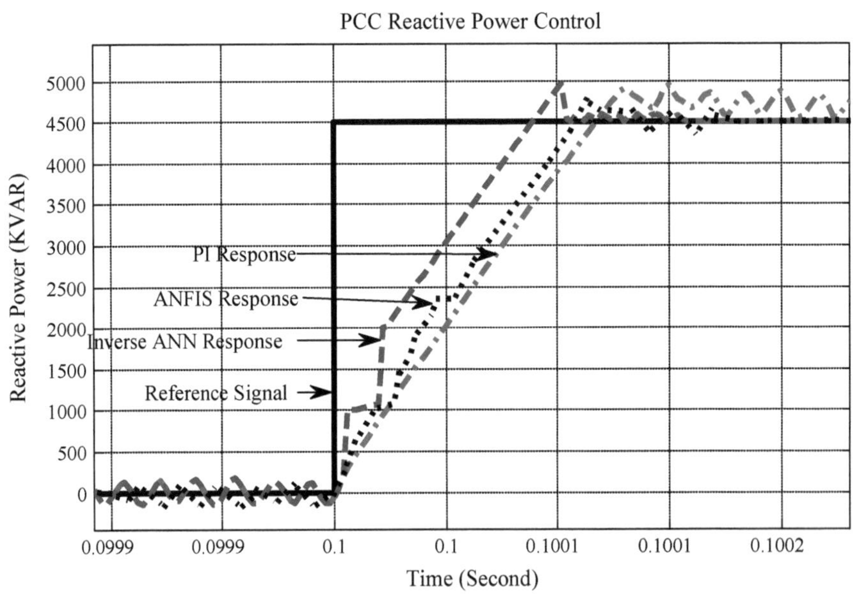

Fig. 22 Comparison of controller performance for reactive power control

that the PI, ANFIS, and inverse ANN controllers perform satisfactorily in regulating the value of active power at PCC. Moreover, the ANFIS and inverse ANN controllers having computational intelligence perform better than the conventional PI controllers. However, the inverse ANN controller is the best among the three in terms of design and performance. The PI and the ANFIS controllers exhibit oscillatory response and have large settling time. However, the response of inverse ANN controller is less oscillatory and reaches the steady state value in lesser time. Figure 22 depicts the performance of controllers for reactive power control at PCC. The reference signal for reactive power was decided, so as to match the power factor requirements according to the grid code. As per the grid code requirements, it is generally desired that the power factor at PCC lies in the range from 0.95 to 1. It can be observed from the response that the reactive power control exhibited by the inverse ANN controller is best as compared with the PI and the ANFIS controller. This is because the settling time in case of inverse ANN controller is minimum; however, the peak overshoot though maximum is tolerable.

6 Conclusion

The PI controllers possess relatively simpler control structure due to their linear nature and ease in hardware implementation. The ANFIS and ANN controllers with computational intelligence (CI) have complex structure and exhibit better dynamic response under wider range of operation. They also exhibit better damping and lower settling time as compared with PI controllers. Hence, the shortcomings of PI controllers can be effectively mitigated with the help of controllers with CI. Moreover, the tuning efforts in the intelligent controllers are quite less vis-à-vis PI controllers. The case study presented in this chapter is sufficient enough to corroborate the superiority of intelligent controllers over the conventional PI controllers. The PI controllers are designed with a linear plant model or by linearization of a nonlinear plant at a particular operating point. When the operating point changes, the performance of PI controller deteriorates, but the nonlinear controllers can perform with accuracy even beyond the range of operation. Due to simplicity in design and training, the inverse ANN controllers are relatively better than their ANFIS counterparts. Hence, in many nonlinear control applications, the inverse ANN controllers can be used satisfactorily. Also, it can be concluded that the controllers with CI have enormous potential for applications in the control of power electronic systems for wind generation purposes.

Appendix

(1) Turbine Parameters: Nominal Power = 30 kW, Base Wind Speed = 12 m/s, Cut-in wind speed = 3 m/s, Cutout wind speed = 14 m/s

(2) Generator Parameters: Nominal Power = 30 kW, Nominal Voltage = 460 V (L-L), frequency = 60 Hz, $J = 1.6$ kg m^2, $P = 2$, $R_s = 0.087\Omega$, $R_r = 0.228$, $L_s = L_r = 0.8$ mH, $L_m = 34.7$ mH.
(3) PI controller parameters: $K_p = 5$, $K_i = 1000$.

References

1. Lin Bor-Ren 1995 Power converter control based on neural and fuzzy methods. Electric Power Systems Research, vol. 35, 960–970.
2. Lofti Kristen, Aribia Ben, Housem, Abdallah Hadj and Abderrazak Ouali 2008 ANN for multi objective reactive compensation of a power system with wind generators. Electric Power System Research, vol. 78, 1511–1519.
3. Ko Hee-Sang, Yoon Gi-Gab, Kyung Nam-Ho and Hong Won-Pyo 2008 Modeling and control of DFIG-based variable speed wind turbine. Electric Power Systems Research, vol. 78, 1841–1849.
4. Jerbi Lilia, Lofti Kristen and Abderrazak Ouali 2009 A fuzzy logic supervisor for active and reactive power control of a variable speed wind energy conversion system associated to a flywheel storage system. Electric Power Systems Research, vol. 79, 919–925.
5. Singh Bharat, Singh S.N. 2010 Intelligent control of power electronic systems for wind turbines. Wind Power Systems Applications of Computational Intelligence, vol. 15, 678–686.
6. Soraes Orlando, Goncalves Henrique, Martins Antonio and Carvalho Adriano 2010 Non-linear control of the doubly fed induction generator in wind power systems. Renewable Energy, vol. 35, 1662–1670.
7. Martinez-Rojas Marcela, Sumper, Andreas, Bellmunt Gomis, Orio, Andreu Sudrià and Antoni 2011 Reactive power dispatch in wind farms using particle swarm optimization technique and feasible solutions search. Applied Energy, vol. 88, 4678–4686.
8. Amimeur H, Aouzellag D, Abdssemed R and Ghedamsi K 2012 Sliding mode control of a dual stator induction generator for wind energy conversion systems. Electrical Power and Energy Systems, vol. 42, 6–70.
9. Tang Yufei, He Haibo, Ni Zhen, Wen Jinyu, Sui Xianchao 2013 Reactive power control of grid connected wind farm based on adaptive dynamic programming. Neuro computing, vol. 125, 125–133.
10. Hamane B., Benghanem M., Bouzid A.M., Belabbes A., Bouhamida M. and Draou A. 2012 Control for variable speed wind turbine driving a doubly fed induction generator using fuzzy PI-control. Energy Procedia, vol. 18, 476–485.
11. Singh Bharat, Singh S.N. 2009 Wind power interconnection into the power system: A review of grid code requirements. The Electricity Journal, vol. 22, 54–63. Wind Power Systems Applications of Computational Intelligence, vol. 15, 678–686.
12. Tsili M. and Papathanassiou S. 2009 A review of grid code technical requirements for wind farms. IET Renew. Power Gener., vol. 3, 308–332.
13. Erlich I. and Bachmann U. 2005 Grid code requirements concerning connection and operation of wind turbines in Germany. IEEE PES General Meeting, vol. 2, 1253–125.
14. Indian wind grid code draft document, www. cwet. tn.
15. Morales A, Robe X, Sala M, Prats P, Aguerri and Torres E 2008 Advanced grid requirements for the integration of wind farms into the Spanish transmission system. IET Renew. Power Gener., vol. 2, 47–59.
16. Arulapalam A, Ramtharan G, Jenkins N, Ramachandramurthy V.K., Ekanayake J.B. and Strbac G 2007 Trends in wind power technology and grid code requirements. In: International Conference on Industrial and information Systems (ICIIS) Srilanka, pp 129–133.

17. Chen Z. and Spooner E. 2001 Grid power quality with variable speed wind turbines. IEEE Trans. Ener. Conv., vol. 16, 148–154.
18. Behera Ranjan and Gao Wenzhong 2010 A novel controller design for grid side converter of doubly fed induction generator for wind power interface and experimental investigation. Electric Power Components and Systems, vol. 38, 1531–1545.
19. Seker Murat, Zergeroglu Erkan and Tatlicioglu Enver 2013 Non-linear control of variable speed wind turbines with permanent magnet synchronous generators: a robust back stepping approach. International Journal of System Science, vol. 2, 1–13.
20. Ahmed Munir, Al-Geelani and Gupta S.P. 2002 Cyclo-converter excited squirrel-cage machine as a wind power converter. Electric Power Components and Systems, vol. 30, 57–76.
21. Melicio R, Mendes and Catalao 2010 Integrated modelling and simulation of a variable speed wind generator. Renewable Energy, vol. 35, 2165–2174.
22. Yazdani Amirnaser and Iravani Reva, 2007 Voltage sourced converters in power system modeling, control and applications. John Wiley and Sons Limited, pp. 450–480.
23. Alexandridis Alex, Stogiannos Marios, Kyriou Alexandra and Sarimveis Haralambos 2013 An offset free neural controller based on a non-extrapolating scheme for approximating the inverse process dynamics. Journal of Process Control, vol. 23, 968–979.
24. Kusagur Ashok, Kokad S.F. and Sankar Ram B.V. 2010 Modeling design and simulation of an adaptive neuro-fuzzy inference system (ANFIS) for speed control of Induction motor. International Journal of Computer Applications, vol. 6, 29–44.
25. www.mathworks.com.
26. Sumathi S, Kumar Ashok and Surekha P. 2015 Solar PV and Wind Energy Conversion Systems. Springer, pp 309–390.
27. Ahmed Munir 2010 Integrated modeling and simulation of a variable speed wind generator. Renewable Energy, vol. 35, 2165–2174.

Power Flow Constrained Short-Term Scheduling of CHP Units

Manijeh Alipour, Kazem Zare and Heresh Seyedi

Abstract The electric power system consists of units for electricity production, devices that make use of the electricity, and a power grid that connects them. The aim of the power grid utilities is to enable the reliable transportation of electrical energy from the production to the consumption, while satisfying system constraints, and all these for the lowest possible price. Conventional power system is facing the problems of gradual depletion of fossil fuel resources, poor energy efficiency, and negative environmental effects. These problems have persuaded system utilities to a new trend of power generation. The new trend incorporates power production at distribution voltage level by using non-conventional or renewable energy sources such as natural gas, biogas, wind power, solar photovoltaic cells, fuel cells, combined heat and power (CHP) systems, and micro turbines. Microgrids (MGs) are accounted as the building blocks of the future power systems known as smart girds. This chapter presents the power flow constrained short-term hourly scheduling of DG units. In the most of the MG scheduling literature, the physical constraints of electric power transmission, known as power flow constraints, has not been taken into account. This simplification may result in a solution that is not technically acceptable. In this study, a MG incorporating cogeneration facilities, conventional power units, and heat only units are considered. The optimal scheduling determines the performance of units in order to supply whole electrical and thermal demand of the MG as well as determining the amount of exchanging power between main and microgrid. In addition, the heat–power dual dependency characteristic in different types of CHP units are considered, and all technical constraints of generation units have been satisfied as well. A mixed-integer linear formulation has been employed to model the non-convex feasible operation region of CHP unit. In this study, a heat buffer tank, with the ability of heat storage, has been incorporated in the proposed

M. Alipour (✉) · K. Zare · H. Seyedi
University of Tabriz, P.O. Box: 51666-15813, Tabriz, Iran
e-mail: alipour@tabrizu.ac.ir

K. Zare
e-mail: kazem.zare@tabrizu.ac.ir

H. Seyedi
e-mail: hseyedi@tabrizu.ac.ir

B. Azzopardi (ed.), *Sustainable Development in Energy Systems*,
DOI 10.1007/978-3-319-54808-1_8

framework. Moreover, in order to consider realistic model of the problem, network operation constraints such as voltage magnitude of buses and line flow limits are taken into account.

Keywords Microgrids · Distributed generation (DG) · Short-Term scheduling · Combined heat and power (CHP)

1 Introduction

In recent years, the rapid escalation in implementation of fossil fuel, poor energy efficiency, and environmental issues have been a significant concern for many utilities, modern societies, and researchers [1, 2]. Effectively use of distributed energy resources (DER), (e.g., wind, biomass, solar, and hydro), combined heat and power (CHP) systems and energy storage technologies may result in a more flexibility, low cost, environmentally, friendly and reliable energy [3, 4]. Aggregation of various DERs, storage devices, and loads, at medium and low voltage distribution systems, forms small power systems called microgrids (MGs). The MGs can be utilized either interconnected or isolated from the main distribution grid as a controlled entity [1, 5].

Recently, incorporating CHP units in MGs have drawn more attention [6]. Utilizing the CHP systems for simultaneous supply of electrical and thermal energy is the primary motivation. In addition, conversion of primary fossil fuels, even by the most modern combined cycle plants, could only achieve efficiencies between 50% and 60%. However, during electricity generation process of CHP units employing waste heat to provide thermal energy will result in fuel conversion efficiencies of the order of 90% [7]. It should be mentioned that the heat and power outputs of CHP units are non-separable and dependent to each other. In other words, in a CHP system, the power generation limitations depend on the heat generation of system, and the heat generation borders depend upon the power generation of the system. In [8, 9], the CHP economic dispatch (CHPED) problem is handled envisaging heat–power dependency feature.

In deregulation and restructured power system, the MG owner tries to supply the MG electrical and heat demand at minimum cost. In this regard, the MG owner would use various resources such as self-generating facilities and highly competitive electricity markets. Due to MGs major technological and regulatory innovation of small-scale on-site CHP-based DERs, MG has become empowering to compete with traditional centralized electricity plant [10]. However, the CHP-based MG owner should consider network operation constraints as well as all technical constraints of generation units in the scheduling problem.

Most researchers concentrate on the management and scheduling of MGs [11–14]. In [11], an intelligent MG energy management with aim of emission and operation cost minimization has been investigated. In [12], a model to make energy trading decisions has been proposed. The proposed model in [12] determines scheduling of units envisaging systems constraints.

The scheduling problem of a MG with islanding constraints is studied in [13]. The objective of [13] is to minimize the MG total operation cost. The problem is decomposed into a grid-connected operation master problem and an islanded operation subproblem. A stochastic framework for operation management of MGs has been proposed in [14]. This paper considers the grid-connected mode, which includes photovoltaic panel (PV), wind turbine (WT), micro-turbine, fuel cell, and energy storage devices.

Operation of CHP units in terms of satisfying both thermal and electrical demand has been investigated in many works [9, 15, 16]. In these works, the hourly scheduling of industrial and commercial customers with cogeneration facilities has been studied, taking the feasible operation region of CHP units into account. In [17], operation of a micro-CHP-based residential MG has been investigated. Thermal load has been studied in [17] in terms of the required hot water and desired building temperature. A mathematical framework for operation of micro-CHPs in a MG has been addressed in [18]. In this work, the grid-connected mode is considered for MG, in which, the MG is able to interchange the electrical energy with the main grid. The objective of the proposed model in [18] is to minimize total costs while meeting heat demand of the grid. In [19], a stochastic programming framework for optimal 24-h scheduling of CHP-based MGs has been proposed. The authors of [19] aim at finding the optimal set points of energy resources for profit maximization, considering uncertainties and demand response programs.

Considering the physical constraints of electric power transmission in the MG scheduling is very vital, especially in the presence of multiple demands. Operational challenges of a MG associated with renewable energy resources (RES) and controllable loads have been addressed in [20]. The power flow (PF) constraints have been taken into account in [20]. Power flow constraints in MG scheduling problem with multiple demands have not been taken into account in all above mentioned works.

In the current chapter, 24-h PF-constrained scheduling of CHP-based MG is conducted. The purpose of the work is to take advantage of the opportunity, to sell any excess electricity to the market in order to maximize the revenue regarding to the prices in the day-ahead market. The unit's operating costs as well as start-up and shutdown costs and cost of power purchases from the power market have been taken into account while satisfying the heat and power demand of the MG. Moreover, network operation constraints such as voltage magnitude of buses and line flow limits have been considered to simulate more realistic model of the problem.

This chapter assumes that the MG possesses a power-only unit, a boiler unit, and two cogeneration facilities. In addition, a heat buffer tank, with the ability of heat storage, has been incorporated in the proposed model. The solution of the scheduling problem meets the terms of technical constraints of all facilities, comprising of minimum up/down time of the facilities, minimum and maximum capacity of units, and dual dependencies of heat and power generation in the CHP systems. The detailed descriptions of proposed model are provided in the following sections.

2 Problem Statement and Mathematical Formulation

The optimal load flow constrained scheduling problem is formulated as a mixed-integer nonlinear programming optimization problem. The objective of the optimal operational scheduling of the CHP-based MG is minimizing the cost of power and heat procurement over a day-ahead period of time, in the presence of several constraints.

2.1 *Objective Function*

The objective function of the scheduling problem is to minimize the total cost. The facilities' operation cost is of importance in the MG scheduling problem. Owing to severe problems of frequent turning on and off [18], the diminishing of units' start-ups and shutdowns is indispensable. Hence, against the most existing works, the all units' start-ups and shutdowns are incorporated in the objective function. It is noteworthy to say that it is assumed that the MG is utilized in grid-connected mode that can buy/sell the electricity from/to the main grid regarding to the market prices.

It is assumed that the CHP-based MG is equipped with conventional power and heat-only units, CHP units, and the heat buffer tank. Therefore, the objective function in the load flow constrained self-scheduling problem of MG to be minimized encompasses the units' start-up and shutdown costs, units' operational cost, expense of procuring energy from the utility as well as taking advantage of selling power to the market:

$$\begin{aligned} \mathrm{OF} = \sum_{t=1}^{24} \Bigg\{ & (P_t^{\mathrm{grid}} \times \lambda_t) + \sum_{k=1}^{N_{\mathrm{CHP}}} C_k(P^{\mathrm{CHP}}, H^{\mathrm{CHP}}) + \sum_{l=1}^{N_P} C_l(P^P) + \sum_{m=1}^{N_b} C_m(H^b) \\ & + \sum_{h \in k,l,m} \mathrm{CSU}_{h,t} \times \mathrm{SU}_{h,t} + \mathrm{CSD}_{h,t} \times \mathrm{SD}_{h,t}) \Bigg\} \quad h \in k,l,m \end{aligned} \tag{1}$$

where k, l and m represent the indices for cogeneration units, power-only units and boiler units, respectively. P_t^{grid} is the amount of electricity sold to the main network, and λ_t indicates the forecasted market price at time t. Also, $\mathrm{CSU}_{h,t}/\mathrm{CSD}_{h,t}$ stand for the start-up/shutdown cost and $\mathrm{SU}_{h,t}/\mathrm{SD}_{h,t}$ show the binary variables representing the start-up/shutdown status of hth system at time interval t. C_{ht} is the total operation cost of generation facilities, which will be described in the following section.

The total operation cost of a CHP unit [21], conventional power-only, $C_{l,t}^P$, and heat-only, $C_{m,t}^B$, units, respectively can be defined as:

$$\begin{aligned} C_{k,t}^{\mathrm{CHP}} = {} & a_k \times P_{k,t}^{\mathrm{CHP}^2} + b_k \times P_{k,t}^{\mathrm{CHP}} + c_k + d_k \times H_{k,t}^{\mathrm{CHP}^2} + e_k \times H_{k,t}^{\mathrm{CHP}} + f_k \times H_{k,t}^{\mathrm{CHP}} \\ & \times P_{k,t}^{\mathrm{CHP}} \end{aligned} \tag{2}$$

$$C^{P}_{l,t} = \psi_l \times P^{P}_{l,t} \tag{3}$$

$$C^{B}_{m,t} = \psi_m \times \mathrm{H}^{B}_{m,t} \tag{4}$$

The binary variables $\mathrm{SU}_{h,t}$ and $\mathrm{SD}_{h,t}$ are used to model the start-up and shut-down status of the facilities, as follows:

$$\mathrm{SU}_{h,t} = U_{h,t} \times (1 - U_{h,t-1}) \quad h \in k, l, m \tag{5}$$

$$\mathrm{SD}_{h,t} = (1 - U_{h,t}) \times U_{h,t-1} \quad h \in k, l, m \tag{6}$$

where, $U_{h,t}$ is binary variable, which is equal to 1 if the hth generation unit is selected at time interval t; otherwise it would be zero. a_k, b_k, c_k, d_k, e_k and f_k represent cost coefficient of cogeneration facility, ψ_l and ψ_m indicate the linear cost coefficient of power-only and heat-only facilities, respectively.

2.2 *Constraints*

The objective function is restricted by equality and inequality constraints, which are as follows.

2.2.1 Load Flow Equations

This chapter considers PF equations in the CHP-based MG scheduling problem in order to simulate more realistic framework of the problem. The following equations characterize the flow of power throughout the system, which are determined by Kirchhoff's laws:

$$P^{\mathrm{grid}}_{t} + P^{g}_{i,t} \times U_{i,t} - P^{l}_{i,t} = \sum_{j=1}^{N_{\mathrm{bus}}} \left(|V_{i,t}||V_{j,t}||Y_{ij}| \cos(\theta_{ij,t} - \delta_{i,t} + \delta_{j,t}) \right) \tag{7}$$

$$Q^{\mathrm{grid}}_{t} + Q^{g}_{i,t} \times U_{i,t} - Q^{l}_{t} = -\sum_{j=1}^{N_{\mathrm{bus}}} \left(|V_{i,t}||V_{j,t}||Y_{ij}| \sin(\theta_{ij,t} - \delta_{i,t} + \delta_{j,t}) \right) \tag{8}$$

which are the active and reactive power flow equations, respectively. N_{bus} is number of buses of the MG. Also, $P^{g}_{i,t}$ and $Q^{g}_{i,t}$ are active and reactive power flow of DERs located on bus i, respectively. P^{grid}_{t} and Q^{grid}_{t} stand for active and reactive power bought from the utility through the bus which is connected to the main grid at time t, respectively. $V_{i,t}$ is the voltage of bus i at time interval t. Y_{ij} and $\theta_{ij,t}$ are magnitude and phase angle of feeder's admittance. $P_{i,t}^{l}$ and $Q_{i,t}^{l}$ are active and reactive load of bus i at time t, respectively.

(a) **Voltage limits**

Voltage limits refer to the requirement for the system bus voltages magnitude, $V_{i,t}$, to be kept at permissible range. Moreover, the voltage magnitude for substation buses, V_s, should be maintained at the nominal value V_s^n:

$$V_{\min} \leq |V_{i,t}| \leq |V_{\max}| \tag{9}$$

$$|V_s| = V_s^n. \tag{10}$$

(b) **Exchangeable power limit**

Exchangeable apparent power with the main grid has to be in a limited bound in order to have the stable operation [22].

$$\sqrt{P_i^{\text{grid}^2} + Q_i^{\text{grid}^2}} \leq S_{\max}^{\text{grid}} \tag{11}$$

(c) **Apparent power *Flow limits for branches***

It is essential to keep the apparent power flowing from each branch, $S_{\text{br},t}$ of MG in its admissible range:

$$\sqrt{P_{\text{br},t}^2 + Q_{\text{br},t}^2} \leq S_{\max,\text{br}} \tag{12}$$

2.2.2 Generation Units Constrains

The operating constraints of generation units contain the minimum up/down time limits, ramping rate constrains, and the generation capacity of facilities. The time duration for which the generation units are on/off at time t, $X_{h,t}^{\text{on}}, X_{h,t}^{\text{off}}$, could be expressed as follows:

$$X_{h,t}^{\text{on}} = (X_{h,t-1}^{\text{on}} + 1) \times U_{h,t-1} \times U_{h,t} + (1 - U_{h,t-1}) \times U_{h,t} \quad h \in k, l, m \tag{13}$$

$$X_{h,t}^{\text{off}} = (X_{h,t-1}^{\text{off}} + 1) \times (1 - U_{h,t-1}) \times (1 - U_{h,t}) + U_{h,t-1} \times (1 - U_{h,t}) \quad h \in k, l, m \tag{14}$$

where, $U_{h,t}$ is a binary variable defining on/ off status of the hth generation unit. Minimum up/down time (UT/DT) limits are imposed by:

$$(X_{h,t-1}^{\text{on}} - \text{UT}) \times (U_{h,t-1} - U_{h,t}) \geq 0 \quad h \in k, l, m \tag{15}$$

$$(X_{h,t}^{\text{off}} - \text{DT}) \times (U_{h,t} - U_{h,t-1}) \geq 0 \quad h \in k, l, m \tag{16}$$

The following equations formulate the ramping up/down rate limits ($R^{\text{up}}/R^{\text{down}}$) of power generation and CHP systems:

$$P_{h,t+1} - P_{h,t} \leq R^{\text{up}} \quad h \in k, l \tag{17}$$

$$P_{h,t} - P_{h,t+1} \leq R^{\text{down}} \quad h \in k, l \tag{18}$$

(a) **CHP units**

It is noteworthy to say that the power and heat productions of the CHP units are mutually dependent and could not be organized separately. Two types of feasible operating region (FOR) can be defined for CHP systems [23]. The first and second types FOR of a CHP unit are exposed in Fig. 1. The first type FOR may be characterized utilizing linear formulation and Eqs. (19)–(23) model this FOR in the CHP-based MG scheduling problem [24].

$$P_{k,t}^{\text{CHP}} - P_{k,\text{A}}^{\text{CHP}} - \frac{P_{k,\text{A}}^{\text{CHP}} - P_{k,\text{B}}^{\text{CHP}}}{H_{k,\text{A}}^{\text{CHP}} - H_{k,\text{B}}^{\text{CHP}}} \left(H_{k,t}^{\text{CHP}} - H_{k,\text{A}}^{\text{CHP}} \right) \leq 0 \tag{19}$$

$$P_{k,t}^{\text{CHP}} - P_{k,\text{B}}^{\text{CHP}} - \frac{P_{k,\text{B}}^{\text{CHP}} - P_{k,\text{C}}^{\text{CHP}}}{H_{k,\text{B}}^{\text{CHP}} - H_{k,\text{C}}^{\text{CHP}}} (H_{k,t}^{\text{CHP}} - H_{k,\text{B}}^{\text{CHP}}) \geq -(1 - U_{k,t}) \times M \tag{20}$$

$$P_{k,t}^{\text{CHP}} - P_{k,\text{C}}^{\text{CHP}} - \frac{P_{k,\text{C}}^{\text{CHP}} - P_{k,\text{D}}^{\text{CHP}}}{H_{k,\text{C}}^{\text{CHP}} - H_{k,\text{D}}^{\text{CHP}}} (H_{k,t}^{\text{CHP}} - H_{k,\text{C}}^{\text{CHP}}) \geq -(1 - U_{k,t}) \times M \tag{21}$$

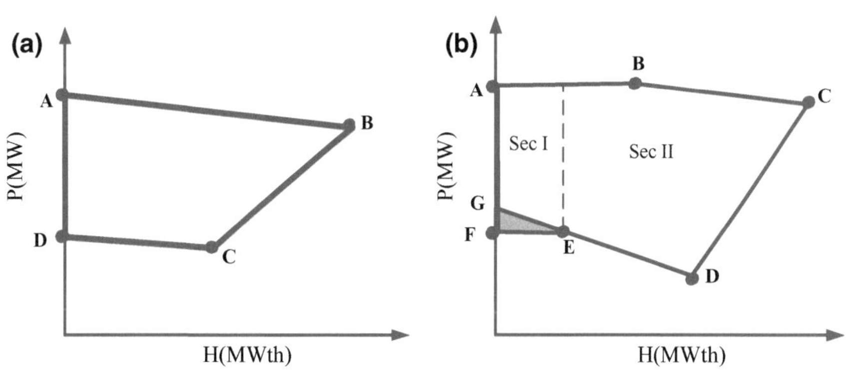

Fig. 1 Power–heat feasible region for a CHP units **a** type 1, **b** type 2

$$0 \leq H_{k,t}^{\text{CHP}} \leq H_{k,\text{B}}^{\text{CHP}} \times U_{k,t} \tag{22}$$

$$0 \leq P_{k,t}^{\text{CHP}} \leq P_{k,\text{A}}^{\text{CHP}} \times U_{k,t} \tag{23}$$

where M is a sufficiently large number, and indices A, B, C, and D represent four bordering points of the FOR in the first type of CHP system. Equation (19) expresses the area under the curve AB. The area upper the curve BC and CD are modeled implementing Eqs. (20) and (21), respectively. Referring to the Eqs. (20)–(21), the output power for a decommitted unit ($U_{K,t} = 0$) would be zero. Moreover, the heat and power generations for a decommitted unit have to be set to zero, which would be executed by Eqs. (22) and (23), respectively.

As can be seen from Fig. 1, the FOR related to type two is non-convex which may be divided into two-convex subregions, namely subregion I and subregion II, as shown in Fig. 1. The type 2 FOR is surrounded by the boundary curve ABCDEFG. Along the boundary curve BC, the power capacity decreases as the heat generation of system increases, while the power capacity declines along the curve CD. In this case, by employing the customary formulation, similar to the first FOR type representation, the colored region (FEG) could not be developed. Therefore, this non-convex area is handled by employing binary variables X_1 and X_2 [23]. Hence, the non-convex region should be divided into two-convex subregions I and II. The subsequent equations model the FOR of CHP unit in the MG scheduling problem, [24]:

$$P_{k,t}^{\text{CHP}} - P_{k,\text{B}}^{\text{CHP}} - \frac{P_{k,\text{B}}^{\text{CHP}} - P_{k,\text{C}}^{\text{CHP}}}{H_{k,\text{B}}^{\text{CHP}} - H_{k,\text{C}}^{\text{CHP}}}(H_{k,t}^{\text{CHP}} - H_{k,\text{B}}^{\text{CHP}}) \leq 0 \tag{24}$$

$$P_{k,t}^{\text{CHP}} - P_{k,\text{C}}^{\text{CHP}} - \frac{P_{k,\text{C}}^{\text{CHP}} - P_{k,\text{D}}^{\text{CHP}}}{H_{k,\text{C}}^{\text{CHP}} - H_{k,\text{D}}^{\text{CHP}}}(H_{k,t}^{\text{CHP}} - H_{k,\text{C}}^{\text{CHP}}) \geq 0 \tag{25}$$

$$P_{k,t}^{\text{CHP}} - P_{k,\text{E}}^{\text{CHP}} - \frac{P_{k,\text{E}}^{\text{CHP}} - P_{k,\text{F}}^{\text{CHP}}}{H_{k,\text{E}}^{\text{CHP}} - H_{k,\text{F}}^{\text{CHP}}}(H_{k,t}^{\text{CHP}} - H_{k,\text{E}}^{\text{CHP}}) \geq -(1 - X_1) \times M \tag{26}$$

$$P_{k,t}^{\text{CHP}} - P_{k,\text{D}}^{\text{CHP}} - \frac{P_{k,\text{D}}^{\text{CHP}} - P_{k,\text{E}}^{\text{CHP}}}{H_{k,\text{D}}^{\text{CHP}} - H_{k,\text{E}}^{\text{CHP}}}(H_{k,t}^{\text{CHP}} - H_{k,\text{D}}^{\text{CHP}}) \geq -(1 - X_2) \times M \tag{27}$$

$$H_{k,t}^{\text{CHP}} \leq H_{k,\text{E}}^{\text{CHP}} \geq -(1 - X_{2t}) \times M \tag{28}$$

$$H_{k,t}^{\text{CHP}} \leq H_{k,\text{E}}^{\text{CHP}} \leq (1 - X_{1t}) \times M \tag{29}$$

$$X_{1t} + X_{2t} = U_{k,t} \tag{30}$$

$$0 \leq H_{k,t}^{\text{CHP}} \leq H_{k,\text{C}}^{\text{CHP}} \times U_{i,t} \tag{31}$$

$$0 \leq P_{k,t}^{\text{CHP}} \leq P_{k,\text{A}}^{\text{CHP}} \times U_{k,t} \tag{32}$$

In the second type, again indexes A, B, C, D, E, and F represent the connection points of the FOR relevant to Fig. 1b. Equations (24) and (25) describe the region under the curve BC and upper the curve CD, respectively. The upper area of the curves EF and DE are described using (26) and (27), respectively. In Eqs. (26–29), CHP unit operation in the first (second) convex subdivision of FOR could be modeled by using $X_{1t} = 1$ ($X_{2t} = 1$). Referring to Eq. (30), the operation region of CHP unit type 2 can be either I or II as the unit status is ON and none of them whenever the unit turns OFF. Furthermore, for a decommited unit, Eqs. (31) and (32) will fix the heat and power production to zero.

(b) **Power-only and heat-only units' constraints**

The capacity restrictions of power and heat-only units can be stated as below:

$$P_l^{P,\min} \times U_{l,t} \leq P_{l,t}^{P} \leq P_l^{P,\max} \times U_{l,t} \tag{33}$$

$$H_m^{b,\min} \times U_{m,t} \leq H_{m,t}^{b} \leq H_m^{b,\max} \times U_{m,t} \tag{34}$$

2.2.3 Heat Buffer Tank

The heat buffer tank has been developed from the model presented in [18]. The heat buffer tank is subjected to the boiler and CHP units. In the proposed system, the heat storing is also possible. The total produced heat, $\overline{H}_t$, could be calculated as:

$$\overline{H}_t = \sum_{k=1}^{N_{\text{CHP}}} H_{k,t}^{\text{CHP}} + \sum_{m=1}^{N_b} H_{m,t}^{b} \tag{35}$$

As the heat exposed to the heat buffer tank is, respectively, effected by the loss (β_{loss}) and extra heat generation (β_{gain}) during shutdown and start-up periods, the real heat, H_t, that the buffer tank would be supplied could be presented as follows [18]:

$$H_t = \overline{H}_t - \beta_{\text{loss}}\text{SU}_{h,t} + \beta_{\text{gain}}\text{SD}_{h,t} \quad h \in k, l, m \tag{36}$$

Hence, the existent heat in the heat buffer tank, B_t, could be stated as:

$$B_t = (1 - \eta)B_{t-1} + H_t - H_t^{\text{D}} \tag{37}$$

where η and H_t^{D} are heat loss rate and heat demand at time interval t (MWth). In addition, the capacity of heat storage is constrained as:

$$B_{\min} \leq B_t \leq B_{\max} \tag{38}$$

In this chapter, the practical state of heat storage system is simulated by envisaging the ramping up/down rates as below:

$$B_t - B_{t-1} \leq B_{\max}^{\text{charge}} \tag{39}$$

$$B_{t-1} - B_t \leq B_{\max}^{\text{discharge}}. \tag{40}$$

3 Simulation Studies

In this section, at first the structure of the considered MG is introduced, and afterward the simulation results of optimal PF-constrained scheduling are presented.

3.1 Microgrid Structure

To investigate the validity and outperformance of the proposed scheduling model, a six-bus meshed MG has been implemented as the test bed here. The single line diagram of this system is depicted in Fig. 2. In the studied case, bus 1 is connected to the main grid, and the MG is able to procure the power from the grid according to the pool market prices. Referring to Fig. 2, the studied MG comprises a power-only

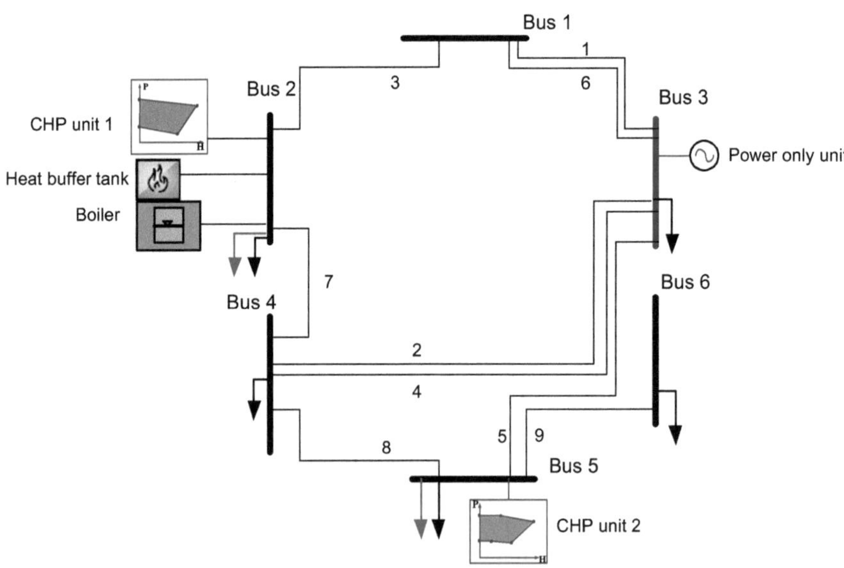

Fig. 2 Single line diagram of six-bus meshed MG

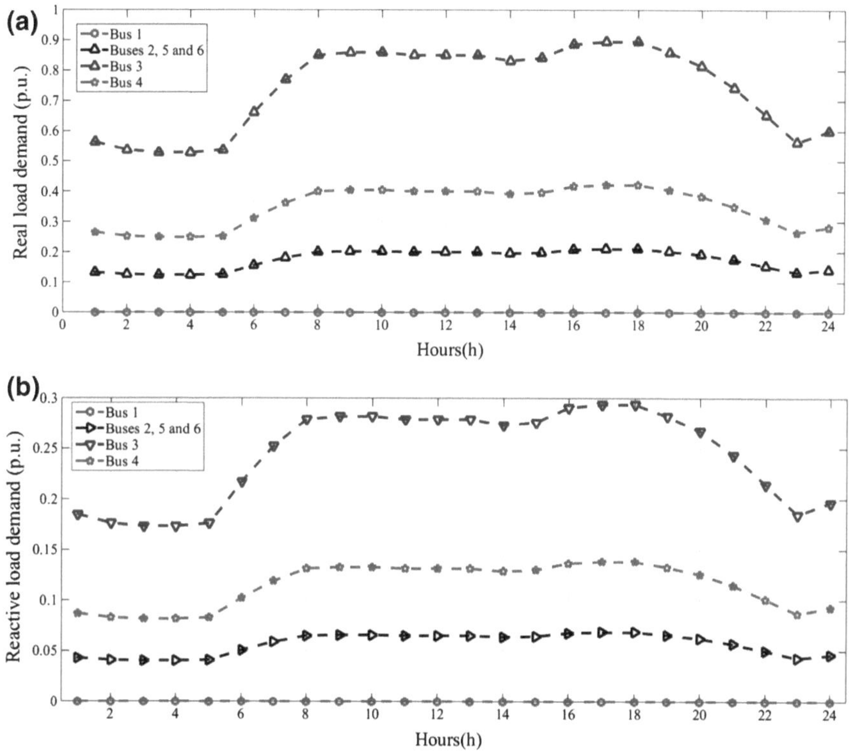

Fig. 3 Bus data **a** active loads, **b** reactive loads

Table 1 Line data (in per unit)

Line no.	Start bus	End bus	R	X	B
1	1	3	0.0342	0.18	0.0106
2	3	4	0.114	0.6	0.0352
3	1	2	0.0912	0.48	0.0282
4	3	4	0.0228	0.12	0.0071
5	3	5	0.0228	0.12	0.0071
6	1	3	0.0342	0.18	0.0106
7	2	4	0.114	0.60	0.0352
8	4	5	0.0228	0.12	0.0071
9	5	6	0.0228	0.12	0.0071

unit and a boiler unit, two cogeneration units with different FORs, and a heat buffer tank along with the electrical and thermal loads. The location of all units is illustrated in Fig. 2. The fundamental network data including the active and reactive loads for all buses and impedance of branches are presented in Fig. 3 and Table 1, respectively [19, 25]. The admissible range for the voltage magnitudes of all buses has been determined to be between 0.95 and 1.05 p.u, respectively. The heat demand of MG is developed from [19, 25] and provided in Table 2. In addition, Table 2 presents the hourly market price in \$/kW. The FOR of cogeneration units is portrayed in Fig. 4. Characteristics of the heat buffer tank are provided in Table 3. The start-up and shutdown cost of units are presented in Table 4. Table 5 provides the cost function coefficients of CHP systems. The cost functions of power-only and

Table 2 Hourly thermal load of the MG and market prices

Hour (h)	Thermal load at bus 2 and 5 (p.u.)	Market price (\$/kWh)	Hour (h)	Thermal load at bus 2 and 5 (p.u.)	Market price (\$/kWh)
1	0.211	0.2704	13	0.324	0.4331
2	0.223	0.2662	14	0.249	0.3866
3	0.172	0.2036	15	0.359	0.6114
4	0.090	0.2178	16	0.383	0.6965
5	0.128	0.2565	17	0.276	0.6178
6	0.264	0.3174	18	0.357	0.4217
7	0.282	0.4466	19	0.316	0.4092
8	0.305	0.4789	20	0.372	0.5974
9	0.265	0.4975	21	0.405	0.4866
10	0.274	0.5427	22	0.404	0.3494
11	0.340	0.5145	23	0.333	0.4368
12	0.261	0.4008	24	0.260	0.4075

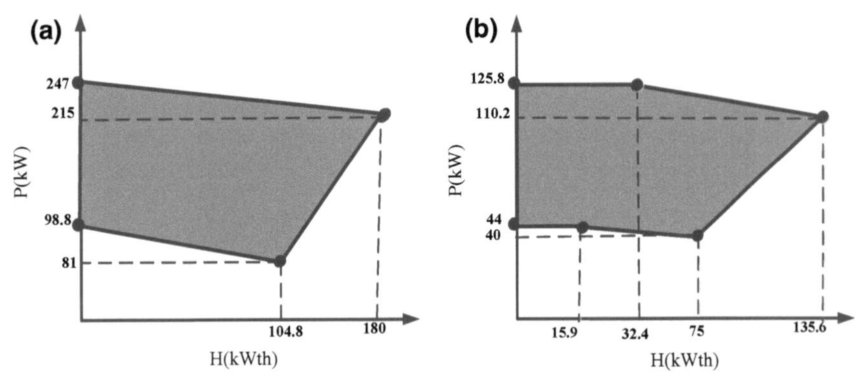

Fig. 4 Power–heat feasible region for CHP units **a** unit 1, **b** unit 2

Table 3 Characteristics of the heat buffer tank

β_{gain}	β_{loss}	η	B_{max}^{charge}	$B_{max}^{discharge}$	B_{max}	B_{min}
0.3	0.6	1%	5	5	30	0

Table 4 Operational constraints and economic data of generation units

Unit/characteristic	R^{up} (kW)	R^{down} (kW)	UT (h)	DT (h)	CSU	CSD
CHP unit 1	100	100	2	2	10	10
CHP unit 2	60	50	2	2	10	10
Power-only unit	20	20	2	2	6	6
Boiler	–	–	–	–	4	4

Table 5 Cost function coefficients of CHP units

Unit	*a*	*b*	*c*	*d*	*e*	*f*
CHP unit 1	0.0345	14.5	2.65	0.03	4.2	0.031
CHP unit 2	0.0435	36	1.25	0.027	0.6	0.011

heat-only are supposed to be linear and stated in Eqs. (41)–(42), respectively. The proposed optimization problem is a mixed-integer nonlinear programming (MINLP) that considers the load flow constraint. Hence, the problem can be classified as an optimal power flow problem, focusing on the energy resource management in the MG. Mathematical modeling of the load flow constrained CHP-based MG scheduling problem is solved by using SBB/CONOPT solver [26] under General Algebraic Modeling System (GAMS) environment [27]. The SBB is a GAMS solver for MINLP models. It is based on a combination of the standard Branch and Bound method known as mixed-integer linear programming (MILP) and the standard nonlinear programming (NLP) solvers, which are already supported by GAMS.

$$C_{l,t}(P^P) = 0.5 \times P^P_{l,t} \quad 0 \leq P^P_{l,t} \leq 150 \text{ kW} \tag{41}$$

$$C_{m,t}(H^b) = 0.234 \times H^b_{m,t} \quad 0 \leq H^b_{m,t} \leq 200 \text{ kWth} \tag{42}$$

In this chapter, two case studies have been investigated:

Case 1: CHP-based MG scheduling neglecting PF constraints.
Case 2: CHP-based MG scheduling considering PF constraints.

3.2 Simulation Results

3.2.1 Case Study 1

CHP-based MG scheduling In this case, the MG scheduling problem has been solved using proposed framework. All technical and economic constraints excluding PF constraints have been envisaged. Table 6 and Fig. 5 summarize the results of case 1. Regarding Table 6, the cost of MG energy supply would be \$339.298. In addition, the MG revenue from the electricity market participation

Table 6 Summarized simulation results of case studies

	Generation cost	Revenue from the sale of power	Cost of buying power	Value of objective function
Case 1	$339.298	$237.014	$0	$102.284
Case 2	$288.951	$155.652	$0	$133.299

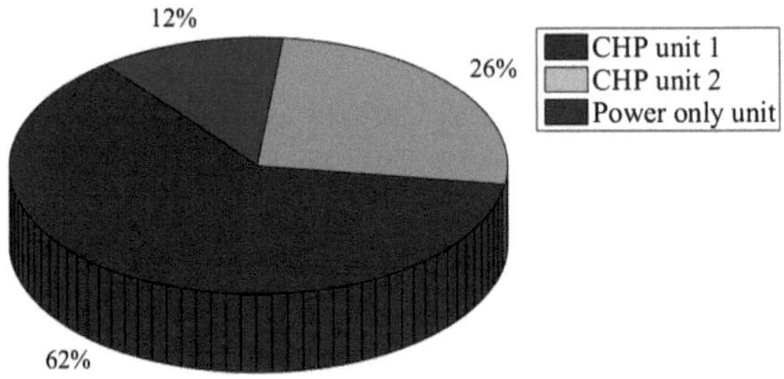

Fig. 5 Percentage of energy supply by different sources

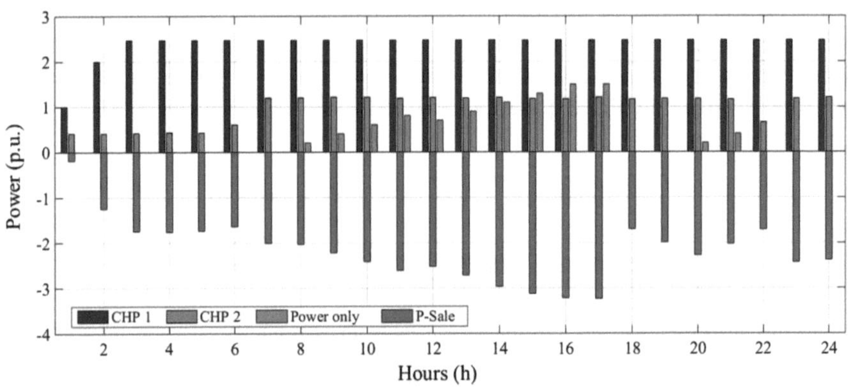

Fig. 6 Generated and interchanged power results of the case 1

over 24-h time horizon would be $237.014. According to Fig. 5, CHP units 1 and 2 will produce about 62 and 26% of total produced power, respectively. Moreover, power-only unit produces only 12% of total generated power. This fact is due to the less operation cost of CHP units, in comparison with power-only unit and also CHP units' heat-power dependency.

Figure 6 illustrates the produced power of all units at the scheduling horizon. As can be seen from Table 6 and Fig. 6, the MG would not purchase power from the market. Due to low operation cost of CHP units, the MG prefers to sell the power to the macrogrid regarding the market prices. It should be mentioned that the amount of exchangeable power with grid has not been limited in this case. Hence, the CHP unit 1 produces power in its maximum capacity at all time intervals, taking into account its ramp up rate constraint in order to take advantages of market participation. The CHP unit 1 does not produce heat at any intervals. Therefore, CHP unit 2 would provide all heat demand of the MG. According to simulation results, the boiler would not participate in supplying heat demand, due to its higher operation cost in comparison with CHP units.

3.2.2 Case Study 2

In the second case, the effect of PF constraints in the CHP-based MG scheduling problem is scrutinized. The MG scheduling problem is solved, considering all technical and economic constraints as well as PF constraints. The results regarding case 2 are provided in Table 6. According to Table 6, the MG revenue from the market participation would be \$155.652. This revenue comes from selling the excess electrical energy to the market as it is operating in the grid-connected mode. The revenue has been decreased about 34.3% in comparison with the case study 1. In addition, the generation cost is decreased to \$288.951. This fact is due to PF constraints which will limit the apparent power flowing from each branch of MG. Moreover, according to Fig. 7, about 7% of produced power will be lost in the branches of network.

Figure 8 presents the voltage magnitude of all buses. According to this figure, the voltage magnitude of all buses is limited between 0.95 and 1.05 p.u. The generated power of units has been depicted in Fig. 9. In this case, again, power-only unit provides power, only in few hours of the day. Referring to Fig. 9 and market prices in Table 2, the MG sells power to the market according to the pool market prices, namely the MG would sell more power in high market price time intervals and less in low price hours to take the most advantages of grid-connected mode and market participation. The heat production of units is portrayed in Fig. 10. As can be seen from Fig. 10, the CHP units produce heat according to their generated power and FOR. The boiler unit does not produce heat at all, due to its high operation cost. Moreover, the buffer tank will be discharged

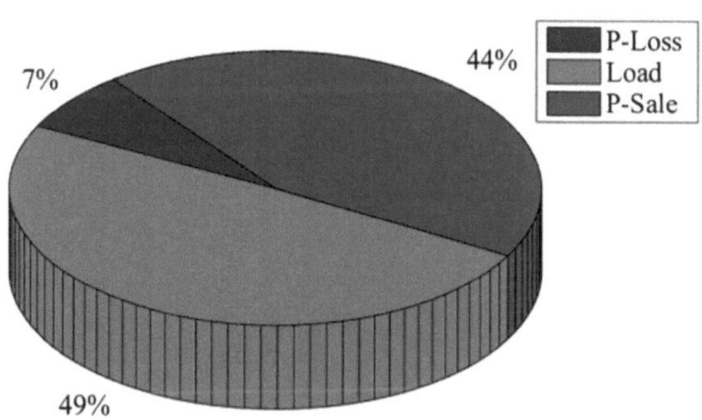

Fig. 7 Percentage of energy spend on various items

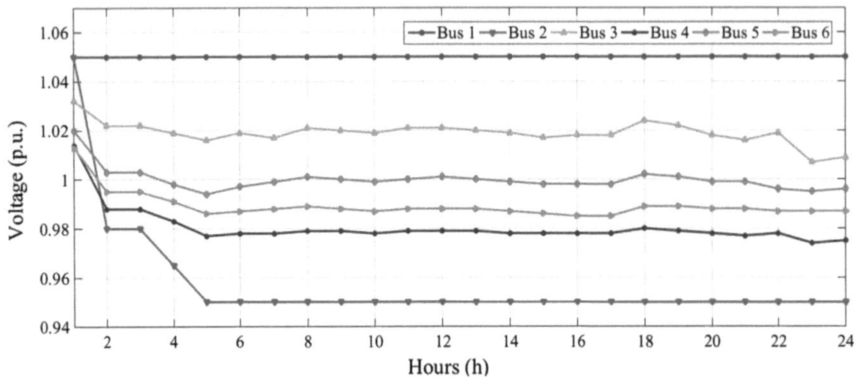

Fig. 8 Voltage magnitude of buses

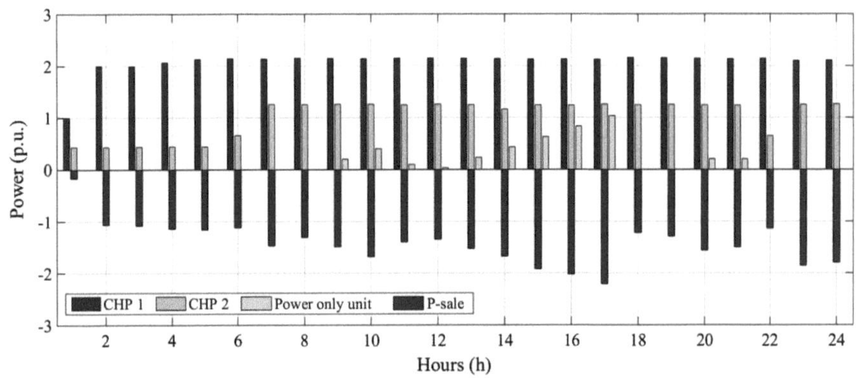

Fig. 9 Generated and interchanged power results of the case 2

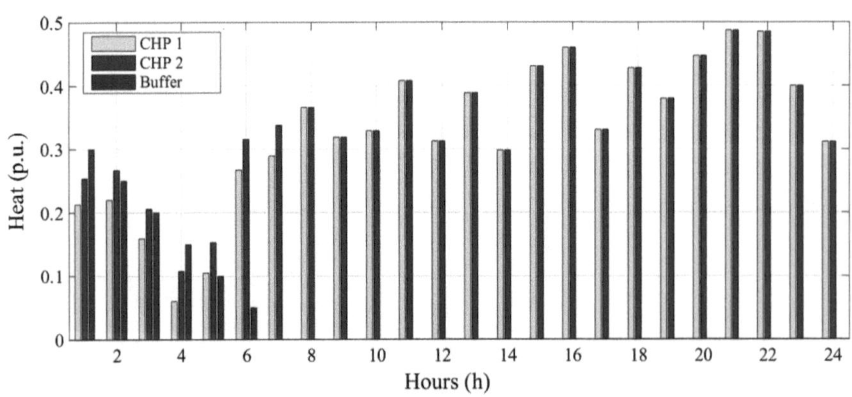

Fig. 10 Generated heat results of the case 2

completely till hour eight, considering its ramp down rate. After hour eight, the produced heat of CHP unit 1 meets the hourly heat demand of MG. Hence, there is no need to store heat in the buffer tank.

4 Conclusions

This chapter presented a power flow constrained programing framework for optimal hourly scheduling of a CHP-based MG, including two conventional power and heat generation units, two types of cogeneration units, and a heat buffer tank. In the optimal scheduling problem of the MG, the objective is to minimize thermal and electrical energy supply cost of MG, as well as taking advantage of market participation to sell excess power in high market price hours and purchase it in low price time intervals. In this chapter, the CHP systems with heat–power dual dependency characteristic are modeled using mixed-integer linear programming formulation. This chapter discussed two case studies in order to scrutinize the impact of PF constraints in the optimal hourly scheduling of CHP-based MGs. According to the simulation results, the proposed model can cover the total thermal and electrical demands with respect to economic criteria as well as physical constraints of the network. The results show that by applying PF constraints the value of objective function has been increased about 30% from $102.284 to $133.299. In addition, total amount of sold power to the pool market has been decreased comparing to the base case (case without PF constraint). This fact is due to PF constraints which will impose apparent power flowing from each branch of MG limitation and power losses in the network.

References

1. H. Jiayi, J. Chuanwen, and X. Rong, "A review on distributed energy resources and MicroGrid," *Renewable and Sustainable Energy Reviews,* vol. 12, pp. 2472–2483, 2008.
2. A. Zangeneh, S. Jadid, and A. Rahimi-Kian, "Promotion strategy of clean technologies in distributed generation expansion planning," *Renewable Energy,* vol. 34, pp. 2765–2773, 2009.
3. P. M. Costa, M. A. Matos, and J. P. Lopes, "Regulation of microgeneration and microgrids," *Energy Policy,* vol. 36, pp. 3893–3904, 2008.
4. P. Dondi, D. Bayoumi, C. Haederli, D. Julian, and M. Suter, "Network integration of distributed power generation," *Journal of Power Sources,* vol. 106, pp. 1–9, 2002.
5. A. Soroudi, M. Ehsan, and H. Zareipour, "A practical eco-environmental distribution network planning model including fuel cells and non-renewable distributed energy resources," *Renewable Energy,* vol. 36, pp. 179–188, 2011.
6. M. Motevasel, A. R. Seifi, and T. Niknam, "Multi-objective energy management of CHP (combined heat and power)-based micro-grid," *Energy,* vol. 51, pp. 123–136, 2013.

7. A. Vasebi, M. Fesanghary, and S. Bathaee, "Combined heat and power economic dispatch by harmony search algorithm," *International Journal of Electrical Power & Energy Systems,* vol. 29, pp. 713–719, 2007.
8. M. T. Hagh, S. Teimourzadeh, M. Alipour, and P. Aliasghary, "Improved group search optimization method for solving CHPED in large scale power systems," *Energy Conversion and Management,* vol. 80, pp. 446–456, 2014.
9. M. Alipour, K. Zare, and B. Mohammadi-Ivatloo, "Short-term scheduling of combined heat and power generation units in the presence of demand response programs," *Energy,* vol. 71, pp. 289–301, 2014.
10. A. K. Basu, A. Bhattacharya, S. Chowdhury, and S. Chowdhury, "Planned scheduling for economic power sharing in a CHP-based micro-grid," *Power Systems, IEEE Transactions on,* vol. 27, pp. 30–38, 2012.
11. A. Chaouachi, R. M. Kamel, R. Andoulsi, and K. Nagasaka, "Multiobjective intelligent energy management for a microgrid," *Industrial Electronics, IEEE Transactions on,* vol. 60, pp. 1688–1699, 2013.
12. C. Changsong, D. Shanxu, C. Tao, L. Bangyin, and Y. Jinjun, "Energy trading model for optimal microgrid scheduling based on genetic algorithm," in *Power Electronics and Motion Control Conference, 2009. IPEMC'09. IEEE 6th International*, 2009, pp. 2136–2139.
13. A. Khodaei, "Microgrid optimal scheduling with multi-period islanding constraints," *Power Systems, IEEE Transactions on,* vol. 29, pp. 1383–1392, 2014.
14. S. Mohammadi, S. Soleymani, and B. Mozafari, "Scenario-based stochastic operation management of microgrid including wind, photovoltaic, micro-turbine, fuel cell and energy storage devices," *International Journal of Electrical Power & Energy Systems,* vol. 54, pp. 525–535, 2014.
15. M. Alipour, B. Mohammadi-Ivatloo, and K. Zare, "Stochastic risk-constrained short-term scheduling of industrial cogeneration systems in the presence of demand response programs," *Applied Energy,* vol. 136, pp. 393–404, 2014.
16. M. Alipour, K. Zare, and B. Mohammadi-Ivatloo, "Optimal risk-constrained participation of industrial cogeneration systems in the day-ahead energy markets," *Renewable and Sustainable Energy Reviews,* vol. 60, pp. 421–432, 2016.
17. M. Tasdighi, H. Ghasemi, and A. Rahimi-Kian, "Residential microgrid scheduling based on smart meters data and temperature dependent thermal load modeling," *Smart Grid, IEEE Transactions on,* vol. 5, pp. 349–357, 2014.
18. G. M. Kopanos, M. C. Georgiadis, and E. N. Pistikopoulos, "Energy production planning of a network of micro combined heat and power generators," *Applied Energy,* vol. 102, pp. 1522–1534, 2013.
19. M. Alipour, B. Mohammadi-Ivatloo, and K. Zare, "Stochastic Scheduling of Renewable and CHP-Based Microgrids," *Industrial Informatics, IEEE Transactions on,* vol. 11, pp. 1049–1058, 2015.
20. W. Su, J. Wang, and J. Roh, "Stochastic energy scheduling in microgrids with intermittent renewable energy resources," *Smart Grid, IEEE Transactions on,* vol. 5, pp. 1876–1883, 2014.
21. G. Piperagkas, A. Anastasiadis, and N. Hatziargyriou, "Stochastic PSO-based heat and power dispatch under environmental constraints incorporating CHP and wind power units," *Electric Power Systems Research,* vol. 81, pp. 209–218, 2011.
22. J. Aghaei and M.-I. Alizadeh, "Multi-objective self-scheduling of CHP (combined heat and power)-based microgrids considering demand response programs and ESSs (energy storage systems)," *Energy,* vol. 55, pp. 1044–1054, 2013.
23. Z. W. Geem and Y.-H. Cho, "Handling non-convex heat-power feasible region in combined heat and power economic dispatch," *International Journal of Electrical Power & Energy Systems,* vol. 34, pp. 171–173, 2012.

24. M. Moradi-Dalvand, B. Mohammadi-Ivatloo, and M. A. Fotouhi Ghazvini, "Short-Term Scheduling of Microgrid with Renewable sources and Combined Heat and Power," in *Smart microgrids, new advances, Challenges and Opportunities in the actual Power systems*, ed.
25. A. K. Basu, S. Chowdhury, and S. Chowdhury, "Impact of strategic deployment of CHP-based DERs on microgrid reliability," *Power Delivery, IEEE Transactions on,* vol. 25, pp. 1697–1705, 2010.
26. "The gams software website," *Available:* http://www.gams.com/dd/docs/solvers/conopt.pdf, 2012 [Online]
27. D. K. A. Brooke, and A. Meeraus. "Gams users guide", Available: http://www.gams.com/docs/ gams/GAMSUsersGuide.pdf [Online].

Optimal Utilization of Solar Energy Resources in Hungary

Attila Talamon and Bálint Hartmann

Abstract By adopting the 2009/28/EC directive, Hungary has committed to increase the share of renewable sources in its gross final energy consumption mix to 14.65% by 2020. In line with this commitment, in 2011 the Hungarian Ministry of National Development has announced the publication of the final version of the Hungarian Renewable Energy Utilization Action Plan (or National Renewable Action Plan—NREAP) for the period between 2010 and 2020. The goal of the NREAP was to guarantee maximal total social recovery by building on the environmental, economical, social, cultural and geopolitical potential of Hungary. However, as the European Commission has also highlighted in 2013, a change of the attitude is needed in many European countries to foster developments in order to achieve the previously undertaken targets. This situation encouraged the joint work of the Hungarian Ministry of National Development and the Hungarian Academy of Sciences which prepared a revision of the NREAP in order to determine future scenarios to utilize the country's renewable energy potential as efficiently as possible and to reach the 2020 targets. The manuscript presents the continuation of the study, prepared in 2014, focusing on the optimal use of solar energy resources. The aim of the chapter was to achieve maximal possible solar primary energy gain by deploying of either solar photovoltaics or solar collectors on the building stock on the country. Since only 1–2% of the stock is exchanged annually, it is very important to increase the energy efficiency of the existing buildings. The study shows that the building sector is able to deliver large energy consumption and CO2 reduction at low costs. The methodology is based on two main database, the building typology from the EPBD-based (Energy Performance of Buildings Directive 2002/91/EC) Hungary Building Energy Strategy, and the LAU-1 (Local Administrative Units, previous NUTS-4) level solar energy potential assessment, both partly created by the authors. Additionally, statistical data on infrastructure, provided by the Hungarian Central Statistical Office, were used to support the

A. Talamon · B. Hartmann (✉)
Environmental Physics Department, Centre for Energy Research,
Hungarian Academy of Sciences, KFKI Campus, Konkoly-Thege Miklós út 29-33,
Budapest 1121, Hungary
e-mail: hartmann.balint@energia.mta.hu

B. Azzopardi (ed.), *Sustainable Development in Energy Systems*,
DOI 10.1007/978-3-319-54808-1_9

decision making process. The study examines possible maximal solar primary energy gain, the ratio of domestic hot water and electricity needs and possible energy generation on a monthly basis. This chapter provides a methodology and an example, how countries can select the best solutions to benefit the most of their solar resources at low costs.

Keywords Solar photovoltaics · Solar collectors · Energy potential assessment · Building typology

1 Introduction

1.1 Policy Background

Directive 2009/28/EC of the European parliament and of the Council [1] sets mandatory national targets for the overall share of energy from renewable sources in gross final consumption of energy and for the share of energy from renewable sources in transport. Previous communication of the Commission demonstrated that a 20% target for the overall share of energy from renewable sources and a 10% target for energy from renewable sources in transport would be appropriate and achievable objectives.

As defined in Article 4 of [1], each European Member State has to provide a National Renewable Energy Action Plan (NREAP) to the European Commission, detailing projections for renewable energy development up to the year 2020. In line with the 2020 climate and energy package of the European Union, in 2011 the Hungarian Ministry of National Development has announced the publication of the final version of the Hungarian Renewable Energy Utilization Action Plan for the period between 2010 and 2020 [2]. The goal of the NREAP was to guarantee maximal total social recovery by building on the environmental, economical, social, cultural and geopolitical potential of Hungary. Considering solar potential of the country, the NREAP has indicated an economical potential of 65 and 25 PJ/a for heat and electricity generation, respectively. However, as the European Commission has also highlighted in 2013, a change of the attitude was needed in many European countries to foster developments in order to achieve the previously undertaken targets. This situation encouraged the joint work of the Hungarian Ministry of National Development and the Hungarian Academy of Sciences which aimed to prepare a revision of the NREAP in order to determine future scenarios to utilize the country's renewable energy potential as efficiently as possible and to reach the 2020 targets.

A multiple-criteria decision analysis of all potential renewable technologies and their respective fuels served as a basis for the most competitive scenarios. The study focused on five aspects: economy, environment, climate, job creation and innovation. Properly quantified indicators and expert weighting were assigned to these five criteria. The aspect of economy included investment and operation and maintenance (O&M) costs. In the case of investment costs, Hungarian and international values were compared, and learning curves were created for each technology, based on recently fulfilled leading project data to show their future potential as well. The aspect of environment and climate both included the examination of life-cycle physiological effects and greenhouse-gas emissions. The effect on job creation was considered with the number of new jobs proportioned to the amount of energy generation during the operation of the power plant. Finally, the number of related patents was chosen as the traditional measure of innovation. When examining the electricity and heating/cooling sectors, regional potentials (wind, hydro, solar, forestry and agricultural by-products, municipal waste, geothermal energy) and demands (heat demand of family houses, apartment houses and block houses) were assessed. The ranking of the potentials and demands was dynamic, in order to enforce market saturation and potential depletion during the optimization.

Results of the multiple-criteria decision analysis have shown that if not supported by direct subsidies or proper policies, solar-based energy generation technologies are not likely to gain momentum; target potential for heat and electricity generation was 1.4 and 0.03 PJ, respectively.

Fortunately, new building energy legislations are much in favour of utilization of locally available solar energy resources. Stochastically changing energy usage costs of the building sector, aspects of environmental protection, economics, sustainability, significant dependence on imported primary energy resources and increasing emphasis on comfort theory during the design process all justify an increasingly detailed elaboration and continuous update of building energy regulations.

Directive 2006/32/EC of the European Parliament and of the Council [3] prescribes elaboration and periodical revision of national energy efficiency action plans. Member States shall adopt and aim to achieve an overall national indicative energy savings target of 9% for the ninth year of application of the Directive (2016), by decrease of 1% annually, compared to the average of years 2002–2006.

According to the COM/2010/0639 communication from the European Commission [4], there is a need to rebalance energy actions in favour of a demand-driven policy. In particular, the transport and construction industries must pursue an active energy savings policy and diversify towards non-polluting energy sources. To stimulate higher energy savings, the Commission should help create appropriate market conditions.

Directive 2010/31/EU of the European Parliament and of the Council [5] has been implemented in the Hungarian legal system by the 1246/2013 (IV.30.) Governmental Decree. The new requirements of the directive bring a "paradigm shift" not only for building energy design and scaling, but also for the relationship of environmental protection and the building industry. The directive regulates construction of new buildings, but the defined significant renovations are also subject of the policy. Strict dates were set for the implementation of new rules:

"Member States shall ensure that:

(a) by 31 December 2020, all new buildings are nearly zero-energy buildings; and
(b) after 31 December 2018, new buildings occupied and owned by public authorities are nearly zero-energy buildings".

The definition of "nearly zero-energy building" means a building that has a **very high energy performance**. The nearly zero or very low amount of energy required should be covered to a **very significant extent by energy from renewable sources**, including energy from renewable sources produced on-site or nearby.

According to the Hungarian implementation of the Directive, "nearly zero-energy building" is a building, which is realized on cost-optimal level (according to the governmental decree on certification of building energy characteristics) or better, in which at least 25% of annual primary energy consumption is supplied by local or nearby renewable energy sources.

Such buildings can be designed and constructed in various ways, but it is crucially important that policymakers did not only emphasize the previously defined criteria of energy efficiency, but also aspects of reaching the cost-optimal level (according to EU methodology), and minimizing the environmental impact of buildings. International and Hungarian experience have both shown that the goals are feasible by using local energy sources.

The implementation of Directive 2002/91/EC of the European Parliament and of the Council [6] has happened through multiple Hungarian decrees.

- 7/2006 (V. 24.) Minister Without Portfolio Decree on the definition of building energy characteristics (and the modificatory 40/2012 (VIII. 13.) and 20/2014 (III. 7.) Ministry of Interior Decree)
- 176/2008 (VI. 30.) Governmental decree on the energy certification of buildings
- 264/2008 (XI. 6.) Governmental decree on the audit of heat production and air conditioning systems.

The most important among these four is the first decree, which includes the calculation methodology for the evaluation of buildings. The same methodology is used for building certification, the legal framework of which process is regulated by 176/2008 (VI. 30.) Governmental decree.

As the above shown regulations indicate, a change is expected in the sector in the near future, and for this reason, the authors have decided to make another revision of the national goals, aiming to achieve maximal possible solar primary energy gain by deployment of either solar photovoltaics or solar collectors on the country's building stock.

1.2 *Building Stock of Hungary*

Before detailing how the building stock of the country was assessed, a short introduction on the Hungarian administrative system is given.

Hungary is divided into 19 counties. In addition, the capital city of Budapest is independent of any county governments. The counties and Budapest are NUTS 3 administrative units. Currently districts (or ridings) serve as the second-level division of the country. (so-called LAU-1 level.) The 198 districts have replaced the 175 sub regions in 2013. Due to the various population density of Hungary, districts are formed by as few localities as 2 (Debrecen and Hajdúböszörmény districts) to as much as 84 (Zalaegerszeg district). The populations ranges between 9000 and 240,000, and the structure of the building stock also shows huge differences. The latest and most proper picture is given by the 2011 census [7], data of which are cited in the following.

The number of flats and residential resorts was 4,390,302 in 2011, an 8% increase compared to 2001. The rate of the growth has exceed the extremely low rate of the previous decade (5.5%) but was still below the steady growth of the 1960–1990 period, which was characterized by an annual rate of 10%. After 1990, number of new constructions has declined; the new increase in the early 2000s was the results of state funded loan programs. After the peak of 2004, a downward trend can be observed, which was accelerated by the financial crisis of 2008 (see Fig. 1).

Growth of the building sector was also spatially diverse. The most dynamic increase was observed in the capital and in county seats, exceeding 10%. Other cities had an 8% increase (similar to national average), while in the case of municipalities the numbers did not reach 5%. Among the regions (NUTS 2 level), the biggest growth was in Central Hungary and Western Transdanubia. (The growth of Pest county, Győr-Moson-Sopron county and Budapest was 2014 and 10%, respectively.) The slowest growth was observed in Borsod-Abaúj-Zemplén, Nógrád and Békés counties, where the expansion of the building stock was below 2%.

Fig. 1 Construction of new buildings in Hungary (*orange*—great depression, *yellow*—World War II, *brown*—socialist planned economy, *green*—introduction of market economy, *blue*—EU membership, *pink*—financial crisis) (Color figure online)

Table 1 Heating of buildings in Hungary

Settlement type	Central heating	Including district heating	Individual room heating	Total
Capital	573,478	221,380	213,856	787,334
County seats	529,542	249,126	193,804	723,346
County cities	77,049	27,512	29,018	106,067
Other cities	717,906	107,137	477,414	1,195,320
Cities subtotal	*1,324,497*	*383,775*	*700,236*	*2,024,733*
Municipalities	539,032	2,423	561,300	1,100,362
Total	2,437,037	607,578	1,475,392	3,912,429

Of the national building stock, 21% is located in Budapest, 51% in cities and the remaining 29% in municipalities. All settlements are characterized by an increasing amount of uninhabited or only seasonably used flats, reaching 11% in 2011. The key drivers of this process are population decline, unemployment, concentration of enterprises in cities and the increased number of seasonably used flats.

The number of inhabitants per 100 flats was 248, a slight decrease from 2001's level (269). The lowest is the housing density in Budapest (213), and the highest is in municipalities. Among the counties, Pest and Szabolcs-Szatmár-Bereg ranks as the highest, while Budapest, Békés and Csongrád are at the other end of the spectrum.

In the last ten years, a small restructuring of the heating of flats has happened. There was a decrease in the number of flats with individual room heating or district heating, but a 9% increase in the share of central heating systems. The latter type of heating represents almost half of the Hungarian building stock, followed by individual room heating (37.7%) and district heating (15.5%).

District heating systems are concentrated in bigger cities of the country; their share is 28.1% in Budapest and above 30% in county seats, but is below 0.5% in municipalities. Regional differences can also be observed. District heating is more likely to be used in the regions of Central Hungary, Central- and Southern Transdanubia, while individual heating is the most popular solution in the eastern regions. Detailed statistical data are shown in Table 1.

2 Methodology

One of the major challenges of the research work was to create a unified building typology that is able to cover the datasets of previously published important and valuable studies, while also providing a relatively simple overview of the building stock. In the following, three widely used building typologies are introduced.

2.1 Typology of the Hungarian Central Statistical Office

The Hungarian Central Statistical Office (HCSO) publishes detailed information on the building stock of the country [7]. Such databases are compiled for each county, separating settlements by their legal status; thus the building stocks of county seats, cities and municipalities are assessed separately. The database types the buildings according to their construction date and the number of flats, located in them. The exact categorization concerning the construction date is the following: built before 1946, between 1946–1960, 1961–1970, 1971–1980, 1981–1990, 1991–2000, 2001–2005 and 2006–2011. Concerning the number of flats, the following categories are defined: 1, 2, 3, 4, 5–10, 11–20, 21–50 and over 51. Table 2 shows the total building stock of the country, using the typology of the Hungarian Statistical Office.

2.2 Typology Used by the National Building Energy Strategy

A more energy-driven approach was used in the typing, created for the National Building Energy Strategy [8]. Although this typing process has made several assumptions and simplifications, it still resulted in the definition of 15 types. (It is easy to prove that another 3–5 types could have been added, since multiflat buildings are handled as a single category.) These types and their corresponding data are shown in Table 3.

2.3 Typology Used for the Requirements of Nearly Zero-Energy Buildings

In 2012, the Ministry of Interior has ordered a study, serving as the basis for the definition of nearly zero-energy buildings [9]. The study required the creation of a building typology, which was based on actual buildings, selected as typical representations of different construction technologies, construction periods and geometrical parameters. This was necessary in order to properly assess the existing building stock, which could not be done using fictitious, automatically generated geometries, an assumption generally used for new buildings. The examination of superstructures also required knowledge of the actual state of the buildings. During the creation of the typology, Geographic Information System of settlements and districts were used, while database of Google Earth and norc.hu provided a useful support as well. The advantage of latter databases is that they allow the fast overview of large areas and many buildings, enabling to determine and choose representative ones. In the next step, concrete units were selected for each building

Table 2 Building stock of Hungary, HCSO typology, 2011

Flats per building	Before 1946	1946–1960	1961–1970	1971–1980	1981–1990	1991–2000	2001–2005	2006–2011	Total
1	513,993	387,248	383,128	426,586	372,854	208,817	121,140	84,942	2,498,708
2	24,210	9581	10,371	9417	5901	3759	3503	3950	70,692
3	8067	1601	1153	1145	1057	952	1030	1191	16,196
4	5353	1280	2510	2795	1375	891	1635	1802	17,641
5–10	10505	3315	4376	5831	5504	2612	3609	2595	38347
11–20	4735	4435	9099	12,891	8295	1203	1889	1137	43,684
21–50	4105	672	1314	3841	1875	223	883	675	13,588
1–	576	42	234	1356	639	13	218	249	3327
Total	571,544	408,174	412,185	463,862	397,500	218,470	133,907	96,541	2,702,183

Table 3 Building stock of Hungary, NBES typology, 2013

	Name	Construction date	Walling	Number of buildings	Number of flats
Type 1	Family house	−1944	Brick, stone, pise	400,537	400,537
Type 2	Family house	−1944	Brick, stone, pise	269,508	269,508
Type 3	Family house	1945–1979	brick, stone, pise	449,213	449,213
Type 4	Family house	1945–1979	Brick, stone, pise	672,128	672,128
Type 5	Family house	1980–1989	Brick, stone, masonry	378,942	378,942
Type 6	Family house	1993–2001	Brick, stone, masonry	198,938	198,938
Type 7	Family house or row-house	After 2001	Brick, stone, masonry	157,885	157,885
Type 8	Apartment house with 4–9 flats	–2001	Brick, stone, masonry	43,981	258,261
Type 9	Apartment house with 4–9 flats	2001–	Brick, stone, masonry	6285	32,241
Type 10	Apartment house with 10 or more flats	–1944	Brick, stone, masonry	10,819	250,871
Type 11	Apartment house with 10 or more flats	1945–2001	Brick, stone, masonry	16,825	268,386
Type 12	Apartment house with 10 or more flats		Medium or large concrete block, poured concrete	10,575	152,567
Type 13	Apartment house with 10 or more flats	−1979	panel	11,502	324,617
Type 14	Apartment house with 10 or more flats	1980–1989	Panel	9635	225,830
Type 15	Apartment house with 10 or more flats	2001–	Brick, other	3770	115,757
Total				2,640,543	4,155,681

type. Using the available construction plans, construction and slope of the roof, and the maximal useful area for the placement of solar photovoltaic and/or solar thermal units were examined (an example is shown in Fig. 2). Larger roof constructions could be identified using solely GIS, due to their adequate resolution.

Examination of block of flats (so-called large panel system buildings) was carried out in a more detailed way for two reasons. First, these buildings are characterized by a relatively high net energy consumption due to the large number of apartments, while having small roof area. Second, this part of the building stock has been constructed according to uniform plans, and thus they are easily typified.

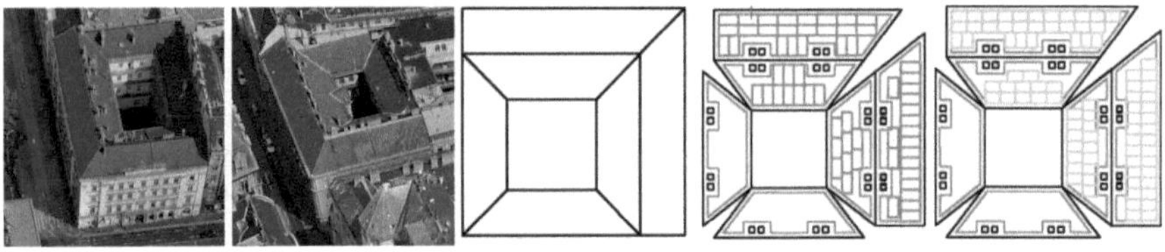

Fig. 2 Assessment of roof areas, and installable solar collectors (*orange*) or photovoltaic panels (*yellow*), [10] (Color figure online)

A total of nine categories (types) were defined; classification was performed based on two set of aspects. The first group includes building properties that can be used to conclude net energy consumption. Energy needs for domestic hot water (DHW) production were defined according to regulations in force (30 kWh/m^2a). Energy needs for heating were determined using structure, construction time and geometrical properties of the buildings. The second group includes properties that are related to the possible use of solar energy. Useful roof area was calculated using the land area covered by the building, after subtracting the areas used by unmovable and permanent objects, like chimney, elevator machine room, corridor, etc. Other objects (vents, antennas) do not play a significant role during the installation, since they can be relocated or bypassed. Of course, individual buildings of the same type may show differences, but since the aim of the work was to determine the magnitudes, such assumptions can be made.

During the typing process, orientation of the buildings was the only examination that was not based on actual buildings, but on five fictitious orientations of the main facade. These five different arrangements are south, southeast, east, northeast and north orientation. Since southwest, west and northwest orientations, considering solar energy gain, are almost the same as southeast, east and northeast orientations, all major possibilities are covered by the selected five arrangements. In the case of centrally symmetric buildings, examination of three orientations is sufficient.

At first, installations of solar collectors on a flat roof may seem an easy task, but surprising experience was gathered during the assessment, and useful areas may prove to be significantly smaller than expected. Roof objects, such as the elevator machine room, not only occupy but also shadow large areas. In the case of flat roofs, solar collectors must be mounted on steel structures to provide proper inclination. Similar loss of roof area is caused by safety regulations, which prescribe leaving at least 1 m between the solar units and the edge of the roof. Proper fixing of the structure must also be guaranteed, and self-shadow effect of the solar collectors has to be taken into consideration in the case of flat roof mounting. Another question is whether to increase area of the collectors (accepting that useful area is lost at times due to self-shadowing) or to avoid self-shadowing during the installation.

During the study, the second approach was chosen, so that none of the collectors are shadowed either by roof objects or by other collectors. (The approximate length of periods without shadowing is 6 h.) To determine maximal energy gain, it was assumed that the collectors are operating all year long. This parameter requires optimization as well; the higher the gain during winter months, the fewer collectors can be installed on the roof.

In the case of high roofs, chimneys, dormer windows and skylights had to be taken into consideration. Even if the examined building had one object on the room, it was assumed to be located on the sunny side, to model worst case scenarios. In addition, 0.5 m safety gap was left between roof objects and the collectors, which is inevitable for the mounting as well. Modification of the roof structure in order to have a better placing for the solar panels was not an option during the study. It also has to be considered that solar collectors and photovoltaic panels are produced in unified sizes, and therefore dead spaces will surely occur during the placing. To minimize this, small panel sizes were chosen.

2.3.1 Comparative Methodology of Energy Needs and Energy Production

During the examination of maximal solar primary energy gains, annual and monthly ratio of DHW and electricity needs and the solar gain was calculated to decide which technology would be more beneficial—solar collectors or solar photovoltaics.

To determine the energy needs of DHW production, 40/2012 (VIII. 13). Ministry of Interior Decree was used. Overproduction of heat energy is considered as a loss in this case. In contrast, electricity needs are handled as theoretically infinite, since legal, technical and economical boundary conditions support the feed in of electricity to the grid. Thus overproduction of electricity is not considered as a loss.

In Hungary, Act LXXXVI. of 2007 and the 273/2007 (X. 19.) Governmental Decree have introduced the definition of small and micro household generation. An electricity generator is considered a micro generator, if:

- it is connected to the public utility grid directly or via a low-voltage line
- nominal capacity of the generator does not exceed maximal permitted consumption of the consumer
- the maximal nominal capacity of the unit is below 50 kVA.

Based on their operation, two types of micro generators can be distinguished. In the first case, all produced electricity is consumed locally and thus there is no injection into the utility grid. In the second case, produced electricity is at least partly injected into the grid. Current legislations allow the customers to use the distribution network as a virtual energy storage, and thus production and consumption periods may differ.

2.3.2 Calculating Primary Energy Needs of Domestic Hot Water Production

To determine the energy needs of DHW production, definitions of the 40/2012 (VIII. 13.) Ministry of Interior Decree were used, as follows:

$$E_{\mathrm{DHW}} = q_{\mathrm{DHW}} \cdot \left(1 + \frac{q_{\mathrm{DHW,v}}}{100} + \frac{q_{\mathrm{DHW,t}}}{100}\right) \cdot \sum (C_{\mathrm{k}} \cdot \alpha_{\mathrm{k}} \cdot e_{\mathrm{DHW}}) + (E_{\mathrm{C}} + E_{\mathrm{K}}) \cdot e_{\mathrm{v}} \quad (1)$$

where

E_{DHW}: primary energy need of domestic hot water [kWh/m^{2}a]
q_{DHW}: net energy need of domestic hot water [kWh/m^{2}a]
$q_{\mathrm{DHW,\ v}}$: loss of domestic hot water distribution [−]
$q_{\mathrm{DHW,\ t}}$: specific losses of storage [−]
C_{k}: capacity factor of domestic hot water production [−]
α_{k}: system proportion [−]
e_{DHW}: primary energy conversion factor [−]
E_{C}: auxiliary energy need of circulation [kWh/m^{2}a]
E_{K}: auxiliary energy need of domestic hot water [kWh/m^{2}a]
e_{v}: primary energy conversion factor (electricity) [−]

This equation is divided into two components, representing the boiler and the solar collector:

$$\begin{aligned} E_{\mathrm{DHW}} = & \left(q_{\mathrm{DHW}} + q_{\mathrm{DHW,v,boiler}} + q_{\mathrm{DHW,t,boiler}}\right) \cdot C_{\mathrm{k,boiler}} \cdot \alpha_{\mathrm{k,boiler}} \cdot e_{\mathrm{DHW,gas}} \\ & + \left(q_{\mathrm{DHW}} + q_{\mathrm{DHW,v,collector}} + q_{\mathrm{DHW,t,collector}}\right) \cdot C_{\mathrm{k,collector}} \cdot \alpha_{\mathrm{k,collector}} \cdot e_{\mathrm{DHW,renewable}} \\ & + (E_{\mathrm{C}} + E_{\mathrm{K}}) \cdot e_{\mathrm{v}} \end{aligned} \quad (2)$$

Net DHW needs are approximately 30 kWh/m^{2}a for both components. Storage losses are equal, since the same storage media is used (the boiler operates as a peak producer), but distribution losses may show small differences. Due to the reason that $e_{\mathrm{renewable}} = 0$, distribution losses of the collector and the effect of capacity factors can be neglected (according to the Decree), and thus (2) can be simplified to the following form:

$$\begin{aligned} E_{\mathrm{DHW}} = & \left(q_{\mathrm{DHW}} + q_{\mathrm{DHW,v,boiler}} + q_{\mathrm{DHW,t,boiler}}\right) \cdot C_{\mathrm{k,boiler}} \cdot \alpha_{\mathrm{k,boiler}} \cdot e_{\mathrm{DHW,gas}} \\ & + 0 + (E_{\mathrm{C}} + E_{\mathrm{K}}) \cdot e_{\mathrm{v}} \end{aligned} \quad (3)$$

A major factor affecting the primary energy need of DHW production is the floor area of the building. According to the regulations of the 40/2012 (VIII. 13.) Ministry of Interior Decree, the bigger the floor area, the less the specific energy need (EDHW).

2.3.3 Energy Gain of Solar Collectors

Energy gain of solar collectors was determined assuming high-quality selective flat plate collectors (Bosch FKT-1S) and vacuum tube collectors (Bosch VK180). Detailed introduction of the arrangement of the collectors on different roof geometries exceeds the scope of present work. The authors have aimed to use maximal roof area, taking into consideration the previously mentioned limitations (objects, safety gaps and shadows). Since the arrangements also determine the number of collectors that can be installed on typified buildings, energy gain can be easily calculated.

In the following, the process is demonstrated on the example of a block of flats building.

Data of the building:

- Orientation: N-S
- Useful room area: 165.8 m^2
- Daily energy need for DHW production: 209 kWh
- Annual energy need for DHW production: 76,285 kWh
- Number of installable collectors: 16

Data of the solar collector:

- Orientation: S
- Inclination: 40°
- Correction factor: 0.98
- Collector type: flat plate
- Absorber area: 2.23 m^2

Daily and monthly per area energy gain was determined, which is used to calculate annual heat energy production of the collector field:

$$Q_{\text{collector annual}} = k \cdot N_{\text{collector}} \cdot A_{\text{absorber}} \cdot \sum Q_{\text{collector monthly}} \tag{4}$$

where

$Q_{\text{collector annual}}$: maximal annual energy production of the collectors [kWh/a]
k: correction factor, indicating the difference to optimal installation (southern facing, 45° inclination)
$N_{\text{collector}}$: number of installable collectors
A_{absorber}: absorber area [m^2]
$\sum Q_{\text{collector monthly}}$: maximal monthly production of a single collector [kWh/m^2month]

Concerning the maximal utilizable irradiation, the results have to be modified based on the actual installation of collectors. In Hungary, the optimal installation for whole year production is southern facing with inclination angles between 40° and 43°.

Table 4 Example for calculating solar energy gains

Month	Solar energy gain of the collector			Heat energy production (kWh)	Monthly energy need of DHW production (kWh)	Solar ratio (%)
	kWh/ (m^2day)	Days in month	kWh/ (m^2month)			
January	0.35	31	10.85	379.4	6479	5.86
February	0.75	28	21.00	734.3	5852	12.55
March	1.35	31	41.85	1463.3	6479	22.59
April	2.05	30	61.50	2150.4	6270	34.30
May	2.35	31	72.85	2547.3	6479	39.32
June	2.65	30	79.50	2779.8	6270	44.34
July	2.85	31	88.35	3089.3	6479	47.68
August	2.85	31	88.35	3089.3	6479	47.86
September	2.55	30	76.50	2674.9	6270	42.66
October	1.65	31	51.15	1788.5	6479	27.61
November	0.65	30	19.50	681.8	6270	10.82
December	0.25	31	7.75	271.0	6479	4.18

Since DHW needs of the building and energy production of the collectors were known, their ratio was calculated, giving the so-called solar ratio. The solar ratio for present example is 28.38%; detailed results are shown in Table 4.

According to the design regulations, solar ratio should reach 70% in Hungary in the case of heat production, and thus the collectors are capable of supplying DHW needs fully during summer months and partially during winter months. Following the practice, the absence of seasonal storage has been taken into consideration during the work, which means that overproduction is lost.

2.3.4 Energy Gain of Solar Photovoltaic Panels

When determining the energy gain of solar photovoltaic panels, the same installation arrangements were used as in the case of the collectors. 240 W polycrystalline modules (Aleo-Bosch S18) were assumed. The PVGIS database, which is based on location, nominal capacity and positioning of the panels, was used to determine the energy gain (number of full load hours). In case one would like to calculate the amount of primary energy that is replaced by the electricity production of the panels, the results have to be multiplied by 2.5. (Primary energy conversion factors are determined by the 40/2012 (VIII. 13.) Ministry of Interior Decree).

2.4 Unification of Building Typologies

As it can be seen from previously detailed building typologies, types and definitions show (in cases significant) difference. Also, the large number of building types makes it more difficult to perform general analysis, without deep understanding of the underlying methodologies. Since the primary aim of the authors was to present results on potential solar energy use, that are easily interpreted by investors and policymakers as well, simplification of the typology was indispensable.

After reviewing the common points of the three typologies, three general groups could be defined, that were present in all of the typologies, with minor differences. These three types are family houses, apartment houses and block of flat buildings (panel houses), which can be usually distinguished, based on their construction date and the number of flats per building.

During our examinations, the following assumptions were used. The type of family houses involves all buildings that have a maximum of 3 flats, regardless of the construction date. This definition involves row-houses (terraced houses) as well, which might be used as apartment buildings, but concerning their geometrical parameters, they are closely related to traditional single family houses. The type of block of flat buildings involves buildings with at least 11 flats, if they were built between 1946 and 1990, but with different shares. For the 1971–1990 period, all such buildings were assumed to be block of flats. For 1961–1970 and 1946–1960, 20 and 15% of such buildings were typed as block of flats. The third type (apartment houses) involved all remaining buildings.

The next step of the process focused on the division of the building stock among LAU-1 level districts. For this, database of the 2011 census was used again (see Sect. 2.1), which included the number of buildings and flats that are located in county seats, cities and municipalities, for all counties and the capital. Assuming that the settlements, belonging in the same category, have similar building stock, the total amount of each building type was divided according to the number of flats in each district. For example, if the county has a total of 100 flats, and 20 of those are located in the examined district, 20% of each building type is assigned to the district. Of course, since districts dominantly include both cities and municipalities, this diversity is also represented after the division of the building stock. After the division, buildings and flats belonging to each of the three building types (family houses, apartment houses and block of flat building) are summed. These values are used to calculate energy needs and solar energy gain.

Considering energy needs, primary energy and electricity are handled differently. Since the National Building Energy Strategy prescribes average specific primary energy needs for the three building types, we used these values. Primary energy need of family houses, apartment houses and block of flat buildings are 384.85, 276.98 and 212.05 kWh/m^{2}a, respectively. (We can see that as the size [volume] of the buildings increases, the relative floor area and thus specific energy needs will decrease.) To determine electricity needs, average district level household consumption of the 2008–2012 period was used; these data are collected by the HCSO. No distinction was made based on wealth of the population, and each flat was assumed to consume the same amount of electricity.

Considering energy production, results of the NZEB typology (Sect. 2.3) were used to determine the maximal installable amount of solar collectors or solar photovoltaic panels, on each of the building types. For solar collectors, annual energy gain was determined based on operational experience. Since overproduction of solar collectors is handled as a loss, as mentioned previously, the bigger the total heat energy need of the building, the higher full load hours were assumed. The exact numbers were determined as 175, 190 and 205 for family houses, apartment houses and block of flat buildings, respectively. Total floor area of the buildings was also taken into consideration, and the average energy production of solar collectors for the three types was calculated as a weighted sum. As a result, annual average solar collector-based energy production of family houses, apartment houses and block of flat buildings was chosen as 527,022,427 and 14,760 kWh/a, respectively, for each building. Knowing the exact building stock of the districts, total energy productions were calculated.

For solar photovoltaic panels, full load hours were determined using the PVGIS database for each district, calculating the average irradiation on panels with east and west orientation. Since the maximal installable power of solar photovoltaic panels has been already determined by the NZEB typology, annual average electricity production of the building types could be calculated. These values show small differences based on the geographical location and the climate characteristics of the districts.

After performing the above detailed calculations, primary energy and electricity needs, and maximal heat energy and electricity production volumes were determined for each building type in all LAU-1 level districts. By comparing these values, primary energy replacement ratio of solar collectors and photovoltaics, and electricity replacement percentages were calculated. In the case of primary energy needs, electricity production of photovoltaic units has a multiplier of 2.5, while in the other two cases, no multiplication is prescribed by the current regulation. The results are detailed in Table 5 (900 full load hours were assumed for solar photovoltaic panels).

3 Results and Discussion

The authors work has included several aspects to analyse the installations possibilities of solar collectors and photovoltaic panels. Not only technical, but also economical and social effects were taken into consideration. Due to space limitations of the format of present publication, only some selected results are detailed, focusing mostly on the aspect of maximizing energy production of solar resources.

3.1 Primary Energy Replacement

As it was detailed Sect. 2.3, according to current regulations, primary energy needs of a building can be replaced by either heat or electricity production; although the replacement factor will be different. In the case of heat energy production, a 1:1

Table 5 Calculation of solar gain for the unified building types

	Number of buildings	Installable solar collector area (m^2/building)	Per building heat energy production (kWh/a)	Average heat energy production of the type (kWh/a)	Installable solar photovoltaic capacity (kW/building)	Per building electricity production (kWh/a)	Average electricity production of the type (kWh/a)
Type 1	400,537	16	2800	Family houses 5270	2.5	2250	Family houses 4320
Type 2	269,508	40	7000		6	5400	
Type 3	449,213	28	4900		4.5	4050	
Type 4	672,128	40	7000		6	5400	
Type 5	378,942	26	4550		4.75	4275	
Type 6	198,938	26	4550		4.75	4275	
Type 7	157,885	28	4900		4.5	4050	
Type 8	43,981	120	22,800	Apartment houses (including Type 15) 22,427	19	17,100	Apartment houses (including Type 15) 16,953
Type 9	6285	120	22,800		19	17,100	
Type 10	10,819	120	22,800		19	17,100	
Type 11	16,825	120	22,800		19	17,100	
	10,575	120	22,800		19	17,100	

(continued)

Table 5 (continued)

	Number of buildings	Installable solar collector area (m^2/building)	Per building heat energy production (kWh/a)	Average heat energy production of the type (kWh/a)	Installable solar photovoltaic capacity (kW/building)	Per building electricity production (kWh/a)	Average electricity production of the type (kWh/a)
Type 12							
Type 13	11,502	72	14,760	Block of flats 14,760	15	13,500	Block of flats 13,500
Type 14	9,635	72	14,760		15	13,500	
Type 15	3770	72	13,680		15	13,500	

replacement factor is assumed, which means that the production of the solar collectors will directly replace heat energy needs of the building. On the other hand, if electricity is produced locally, a 2.5:1 factor has to be used, resulting better replacement factors if solar photovoltaics are used.

Another significant difference between the two technologies is that the production of solar collectors was limited to avoid overproduction (and thus unnecessary losses), while production of solar photovoltaics had no such limitations, and injection of surplus electricity into the grid is an available option. Because of these boundary conditions, replacement factors of solar collectors will be influenced more by the composition of the building stock, and regions with worse solar irradiation characteristics can also perform well from this aspect. In the case of solar photovoltaics, the number of full load hours will have a much bigger effect on replacement factors, but the composition of the building stock is also an important factor.

The spatial distribution of replacement factors for solar collectors and photovoltaics are shown in Figs. 3 and 4, respectively. It can be seen in both figures, that highly urbanized districts (mostly county seats) have the smallest replacement factors around the country. This is the result of the high share of block of flats and apartment houses in the building stock, since the specific roof area of such buildings is significantly lower than of family houses.

In the case of solar collectors, the five worst performing districts are Budapest, Miskolc, Székesfehérvár, Debrecen and Tatabánya, which are all county seats. Interestingly the five best districts are located in the same county: Baktalórántháza,

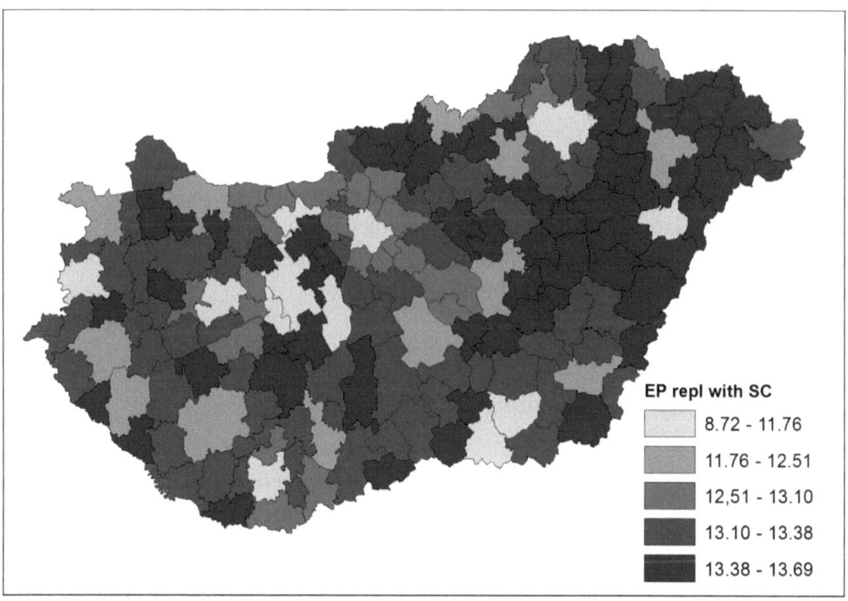

Fig. 3 Spatial distribution of primary energy replacement factors, solar collectors

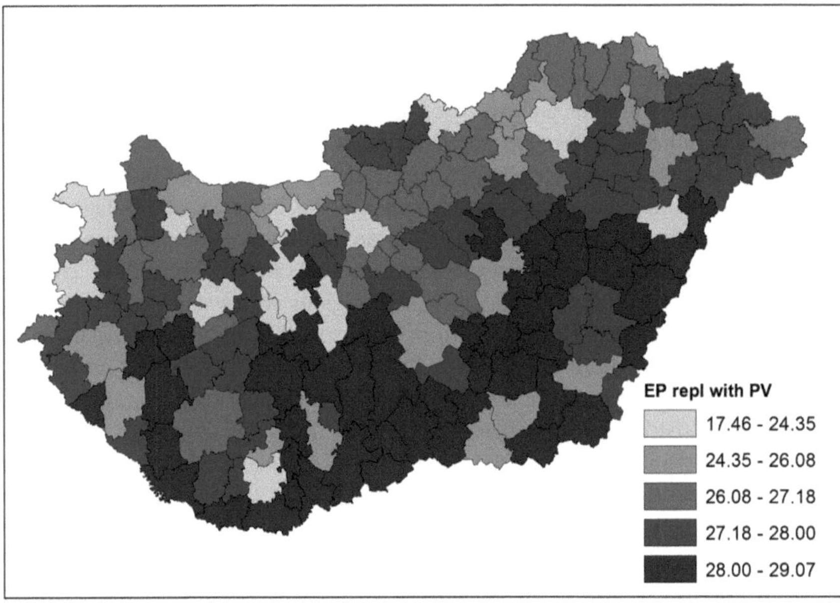

Fig. 4 Spatial distribution of primary energy replacement factors, solar photovoltaics

Csenger, Fehérgyarmat, Nyírbátor and Vásárosnamény are all among the least urbanized areas of the country.

In the case of solar photovoltaics, the worst replacement factors are found in the districts of Budapest, Miskolc, Debrecen, Tatabánya and Veszprém. The top five includes districts of Sellye, Bácsalmás, Baja, Mórahalom and Kunszentmárton.

To highlight the effect of the composition of the building stock, Fig. 5 details how well each building type performed during the analysis. It can be recognized that in general, solar photovoltaics have a higher replacement factor, regardless of the type of the building; this is on one hand the result of the different multiplier, but on the other hand, result of the possibility to inject surplus electricity into the grid. It also can be seen that the highest factors occurred in the case of apartment buildings, but this type shows the highest spread as well.

3.2 Electricity Replacement

The second possibility, that the authors have examined, is the case, when solar photovoltaic production is not considered to be used for the replacement of primary energy, but electricity needs. The major difference is that in this comparison, no multipliers are used in favour of photovoltaics. The spatial distribution of the results (see Fig. 6.) is significantly different than the in the previous cases, shown in

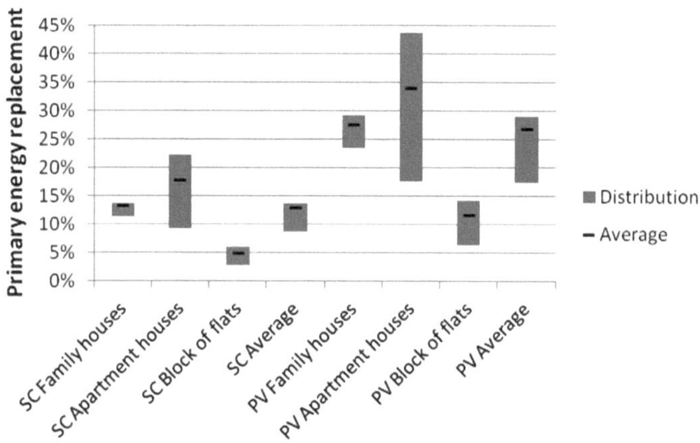

Fig. 5 Primary energy replacement of solar collectors (SC) and photovoltaics (PV) for different building types

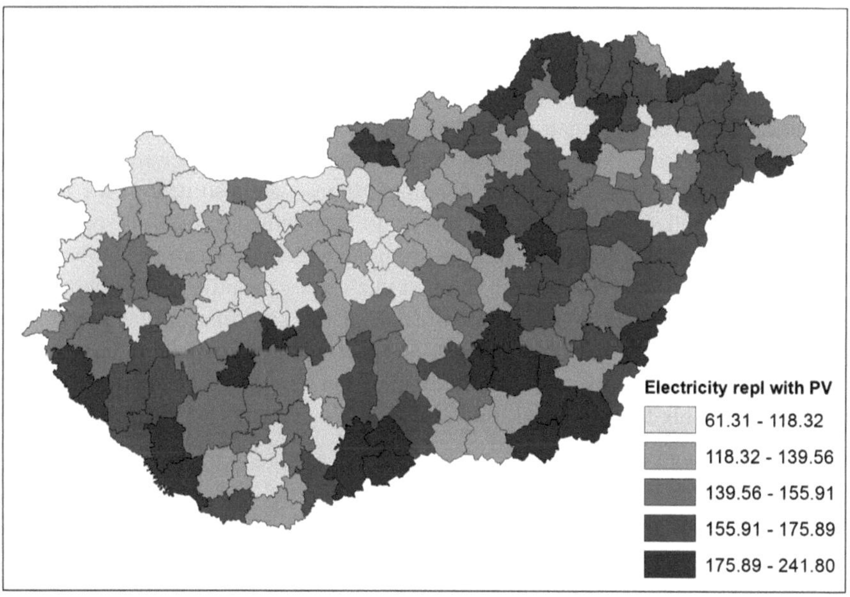

Fig. 6 Spatial distribution of electricity replacement factors, solar photovoltaics

Sect. 3.1. At first, no clear correlation with full load hours can be noticed, since well-performing districts are located in both the southern and northern parts of the country. The underlying characteristic, which is not shown directly by the figure, is the volume of household electricity consumption. Districts with the best replacement factors (Mezőkovácsháza, Cigánd, Tiszaújváros, Enying, and Nagyatád) have

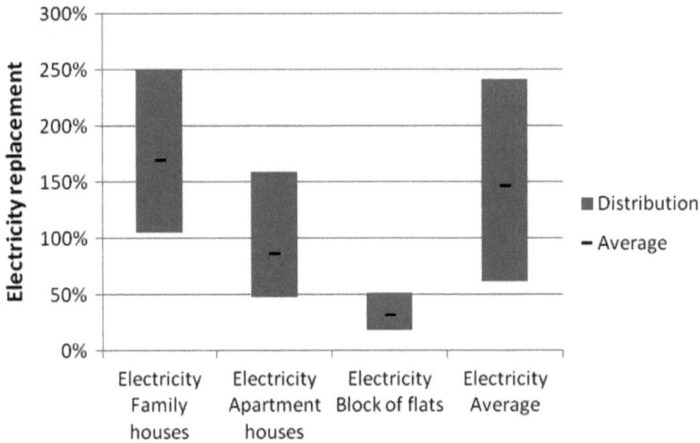

Fig. 7 Electricity replacement of solar photovoltaics for different building types

relatively low per capita electricity consumption. Therefore, if no major difference is observed in the composition of the building stock, such districts will perform well in this aspect. At the other end of the spectrum, districts with the lowest replacement factors are Budapest, Pécs, Debrecen, Oroszlány and Tatabánya; all of them are county seats with the exception of Oroszlány.

Effects of the composition of the building stock are shown in Fig. 7. It can be seen that family houses in all districts of the country could produce more electricity than they consume, and in average, apartment houses are also near to the point of being "self-suppliers". The results of this figure also indicate that if the penetration of solar photovoltaics continues to grow, regions with large share of family houses in their building stock will present the first challenges for utility grid operators, in terms of reverse power flows. However, one of the key aspects of efficient integration of variable renewable energy sources is the so-called system-friendly installation, which assumes that electricity injection into the grid is minimized on an annual level. This technological limitation was also examined by the authors.

3.3 *Primary Energy Replacement with Limitation*

If we assume that with the growth of solar photovoltaics penetration, feed-in possibilities will be limited, the technology will suffer similar overproduction losses as discussed previously in the case of solar collectors. To take this also into consideration, primary energy replacement factors were recalculated for the scenario, when the volume of replaced energy must not exceed the volume of local electricity needs. The results and the comparison with the previous case are shown in Fig. 8. It can be recognized that replacement factors drop significantly, in average from 27 to 17%. The change is mostly caused by family houses, while the other two building types are practically not affected by such change of the regulation.

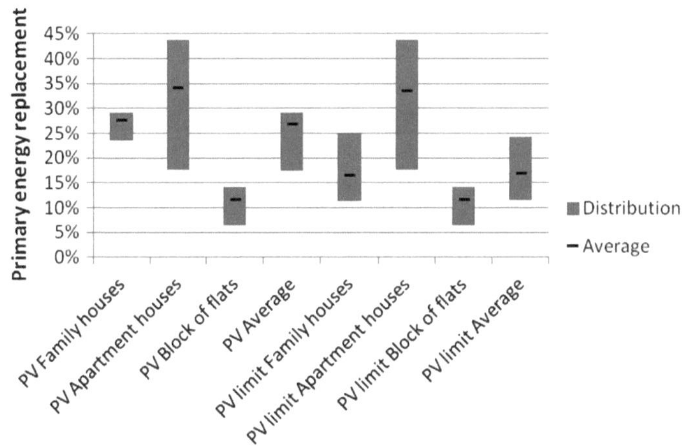

Fig. 8 Primary energy replacement of solar photovoltaics for different building types, with and without limitations

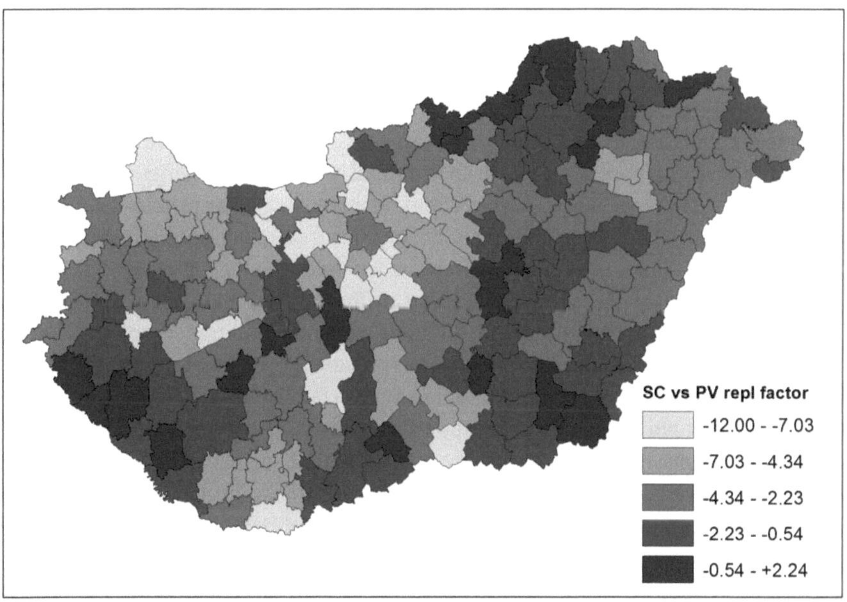

Fig. 9 Primary energy replacement of solar photovoltaics for different building types, with and without limitations

Since this new calculation method (and the concerning regulations) is expected to change the spatial distribution of districts with favourable replacement factors, the authors have performed a comparison of the two scenarios. Figure 9 shows the difference in the primary energy replacement factors between solar collectors and solar photovoltaics. If values are negative, the district is more suitable for deployment of solar photovoltaics, even in the case of possible new regulations. If values are positive or near the break-even, installation of solar collectors can be considered as a real option. The best performing districts are Cigánd, Edelény, Ózd, Tiszaújváros and Enying, several of whom have been already mentioned in this section.

It has to be emphasized, though, that current regulations are much in favour of solar photovoltaics, primarily as the result of the different multiplier of the replacement factor.

4 Conclusions

By adopting the 2009/28/EC directive, Hungary has committed to increase the share of renewable in its gross final energy consumption mix to 14.65% by 2020. So far, very little improvement has been made in the field of small-scale renewable, especially solar energy technologies. Although the revision of the National Renewable Action Plan, carried out by the Hungarian Ministry of National Development and the Hungarian Academy of Sciences, has indicated the need for a change in 2014, clear policies are yet to be defined for the sector. In their absence, investors still find solar technologies favourable, and the penetration has shown an exponential increase in the last couple of years. In late 2015, the authors have prepared a study, which focused on the optimal use of solar energy resources.

The aim of the work was to achieve maximal possible solar primary energy gain by deploying of either solar photovoltaics or solar collectors on the building stock on the country. Since only 1–2% of the stock is exchanged annually, it is very important to increase the energy efficiency of the existing buildings. The methodology was based on the unification of three separate building typologies, all of which are only partly suitable for easily interpretable decision support systems. The unified typology consists of three building types (family houses, apartment houses and block of flat buildings), the corresponding primary energy and electricity needs, and maximal solar energy gains. Using statistical data, provided by the Hungarian Central Statistical Office, LAU-1 level assessment of the building stock was performed, and the potentials of solar collectors and photovoltaics were examined on district level.

Results of the work have shown that in the current regulation environment, installation of solar photovoltaics is more favourable in practically all districts of Hungary, due to the increased multiplier of primary energy replacement. However, if new policies aim to limit reverse power flows in the distribution grid (as a results of solar photovoltaics overproduction), solar collectors will have better performance in the comparison. Results have also highlighted the importance of the proper assessment of the building stock; future work aims to improve the unified typology, to better understand needs and opportunities of Hungarian regions.

References

1. EC, 2009. Directive 2009/28/EC on the promotion of the use of energy from renewable sources and amending and subsequently repealing Directives 2001/77/EC and 2003/30/EC, European Parliament and the Council, 23 April 2009
2. NREAP, 2011. Hungary's Renewable Energy Utilisation Action Plan on trends in the use of renewable energy sources until 2020, Hungarian Ministry of National Development, 2011
3. EC, 2006. Directive 2006/32/EC on energy end-use efficiency and energy services and repealing Council Directive 93/76/EEC, European Parliament and the Council, 5 April 2006
4. EC, 2010a. Communication COM/2010/0639, Energy 2020 A strategy for competitive, sustainable and secure energy, European Commission, 10 November 2010
5. EC, 2010b. Directive 2010/31/EU on the energy performance of buildings, European Parliament and the Council, 19 May 2010
6. EC, 2002. Directive 2002/91/EC on the energy performance of buildings, European Parliament and the Council, 16 December 2002
7. Census, 2011. Population Census 2011, Hungarian Central Statistical Office, 2011. Available online: http://www.ksh.hu/nepszamlalas/
8. NBES, 2015. National Building Energy Strategy, ÉMI Építésügyi Minőségellenőrző és Innovációs Nonprofit Kft., 2015
9. Csoknyai et al., 2012a. Requirements for nearly zero-energy buildings, using renewable energy sources, 2012. (In Hungarian: A megújuló energiaforrásokat alkalmazó közel nulla energiafogyasztású épületek követelményrendszere)
10. Csoknyai et al., 2012b. Building renovations and nearly zero energy use, Magyar Épületgépészet, vol. 61., iss. 7–8., pp. 30-35 (In Hungarian: Épületfelújítások és a közel nulla energiafelhasználás)

Microgrids Operation in Islanded Mode

A.C. Zambroni de Souza, B. De Nadai N., F.M. Portelinha, Jr., Diogo Marujo and D.Q. Oliveira

Abstract The smart grid concept is intended to improve power system operation and control. A feasible path to make the system smarter is through microgrids deployment. A microgrid is a small scale-power system with its own power generation units and deferrable loads, and it may work islanded or connected to the main power grid. The main objective of microgrids in islanded mode is to allow the system to operate even in adverse scenarios, such as faults in main grid, high prices of main grid's power, and supplying remote areas. In the case of an islanding, high priority loads, such as hospitals, transportation and telecommunication facilities must have their supply assured. This is possible due to the penetration of Distributed Energy Resources (DERs), including renewable, fossil, combined heat and power, and energy storage units. However, the operation of microgrids in islanded mode requires more attention due to the higher outage risk since the power generation capacity is limited. Consequently, microgrids may be provided by an Energy Management System (EMS) responsible for managing the scarce power resources to maintain the supply for the highest priority customers connected to the

A.C.Z. de Souza (✉) · B. De Nadai N. · F.M. Portelinha, Jr.
Institute of Electrical Energy, Federal University of Itajuba, Itajuba, Brazil
e-mail: zambroni@unifei.edu.br

B. De Nadai N.
e-mail: nadaibruno@gmail.com.br

F.M. Portelinha, Jr.
e-mail: portelinha@gmail.com

F.M. Portelinha, Jr.
Radiocommunication Reference Center, National Institute of Telecommunications INATEL, Santa Rita Do Sapucaí, Brazil

D. Marujo
Department of Electrical Engineering, Federal University of Technology—Paraná, Medianeira, Brazil
e-mail: diogomarujo@utfpr.edu.br

D.Q. Oliveira
Institute of Electrical Energy, Federal University of Maranhão, Sao Luis, Brazil
e-mail: dq.oliveira@ufma.br

B. Azzopardi (ed.), *Sustainable Development in Energy Systems*,
DOI 10.1007/978-3-319-54808-1_10

grid. Such management strategy is a complicated and ambitious task and demands a robust and reliable communication system. This chapter investigates some control and management issues in microgrids islanded operation mode. Firstly, the main features and requirements of islanded mode in comparison with connected mode are described. Some discussions about control requirements on different control levels are presented. Communications networks are also discussed. These communications networks enable the expected smart features in the microgrids through the bidirectional data flow. The steps for designing a mobile telecommunication network for a microgrid are described, and a study case considering a small microgrid is investigated to show the communication network design steps and the operation of an islanded microgrid during one day.

Keywords Islanded microgrids · Energy management systems · Communication networks · Smart-grids

1 Introduction

Microgrids are a feasible way to deploy the smart grids, since connecting small and smart micro systems in different sites is more realistic and less expensive than building a completely new infrastructure [1, 2]. These distributed microsystems should have their own Distributed Energy Resources (DERs), e.g., wind turbines, photovoltaic arrays, energy storage systems (ESSs), and fossil-fuel microturbines. They also have a communication network, which enables some data exchange between the local controller and Microgrid's Central Controller (MGCC), or between distributed agents. Metering and sensing are also expected features that enable grid smartness.

The microgrid's capacity to operate in islanded mode, the proper operation of the protection schemes and the application of different methodologies of grid reconfiguration enables the self-healing capacity. The continuity of power supply in islanded mode depends on the existence of enough DER installed capacity, proper energy management and other issues related to grid dynamic behavior, e.g. power balancing, low grid inertia, and higher frequency variations.

The islanded operation has some tighter requirements than connected operation. Larger frequency variations, nodal voltage limits, lack of high inertia power sources, and the randomness of the renewable power generation are some issues that should be faced.

This chapter aims to discuss the architecture of a MGCC and the communication infrastructure necessary for its proper operation in islanded mode. The designing process of a mobile telecommunication network for a microgrid is also detailed. Finally, a study case of an islanded microgrid is also presented, where the ideas and concepts previously presented are further investigated.

2 Microgrids Operation in Islanded Mode

One of the desired features of a microgrid is the capacity to operate both in islanded and grid-connected modes. The islanding process occurs by the opening of upstream switches at the substation that interconnects the microgrid and the utility grid. This switching operation may occur due to any intentional operation action or disturbance issues [3].

In an eventual scenario of islanding, the microgrid must be able to supply as many consumers as possible considering safety and quality aspects, i.e., in acceptable ranges of voltage and frequency and assuring a minimal load shed. Thus, the presence of a MGCC able to manage and control all the agents inside the system in islanded mode becomes indispensable.

Some functions of the MGCC should be highlighted, such as [3]:

- To restore voltage and frequency in islanded mode;
- To determine connection and disconnection's policies of loads;
- To optimize the operation based on prices of energy market, weather conditions and load profile;
- To manage the buying and selling of power;
- To guarantee a smooth reconnection and islanding process.

The implementation of a MGCC with all of the previously stated items requires a complex infrastructure regarding data processing and telecommunication networks, enabling a secure and intelligent operation. This differentiates the active distribution system from a microgrid.

In the microgrid, the control is executed by the MGCC and local controllers at the loads and microsources, named here as LC (Load Control) and MC (Microsource Control), respectively [4]. This creates two distinct control approaches: centralized and decentralized.

In the centralized approach, the microgrid is centrally controlled by the MGCC, typically located at the main substation, with a number of functions distributed in a defined hierarchy control.

In the decentralized approach, MC and LC exchange information with the MGCC, providing the set points. The LCs operate considering the priority of load, i.e., if necessary, load shedding is performed in low priority loads while important loads like hospitals, telecommunication and transportation services have maintained the power supply. The MCs control the active and reactive power generation of each micro source based on local information and reference set points sent by MGCC [4].

Both approaches, however, follow a hierarchy defined by the control levels – primary, secondary, and tertiary. This hierarchy proposed in [5] is based on ANSI/ISA-95—an international standard for developing an automated interface between enterprise and control systems. The hierarchy adapted for application in microgrids is made as follows:

- Primary Control: the droop method is used to emulate the behavior of synchronous machines. In this level, the virtual impedance can also be implemented to improve the X/R ratio;
- Secondary Control: guarantees the restoration of frequency and voltage by changing the operation set points of primary control. This level also can include a strategy to resynchronize the microgrid with the utility grid;
- Tertiary Control: controls the power flow between microgrid and utility grid, executed only when microgrid operated in connected mode.

In summary, considering the generation and the adopted hierarchy, the local control is the first level, i.e., the primary control; and the centralized control—the MGCC—is responsible for the secondary and tertiary controls. Figure 1 shows the hierarchy of control in a microgrid. In the pyramid, the bandwidth required to information and data processing issues increases nearer the base.

It is not usual the direct connection of synchronous generators in a microgrid, even they are normally responsible for voltage and frequency control in conventional power systems. For the renewable-based power units, the connection is made by an electronic interface between the power plant and the system, called Power Electronic Converter (PEC) [4]. In this context, the MCs operate the Pulse Width Modulation (PWM) algorithm inside the electronic interface, which is responsible for controlling the active and reactive power injection.

The application of PECs in microgrids constitutes an important issue, since they are responsible for microgrids' operational flexibility, i.e., enabling the transition between connected and islanded mode. In AC microgrids, the PECs, usually identified as inverters, are the final stage , transforming the direct current source into 50Hz alternating current. In general, PECs have two possible operation modes [4, 5]:

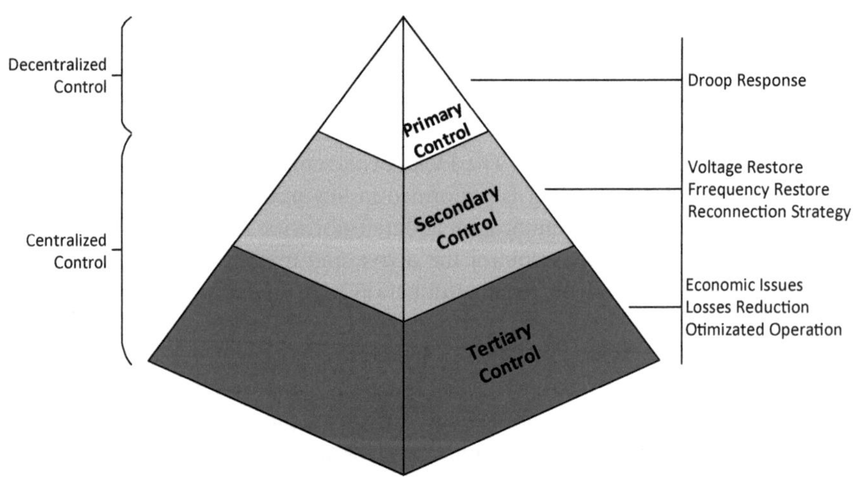

Fig. 1 Hierarchy of control adopted to microgrids

- PQ mode:
  In PQ mode, the PECs operate like current sources, by injecting defined values of active and reactive power. In this case, these sources are not able to control the frequency and the voltage of the system. This kind of converter is synchronized by the PLL (Phase Locked Loop) with the grid's frequency. The set point values can be defined locally by a Maximum Power Point Tracker (MPPT) in the case of photovoltaic models, for example, or by the MGCC. The renewable-based units, which depend on the volatility and intermittence of the weather, cannot be controlled by the MGCC, but are prevalent in PQ mode. In these cases, the power available in the primary source is injected into the grid. PQ converters are generally modeled as "negative load" during power flow simulations they keep injecting constant power regardless the voltage and frequency values [6];
- VSI (Voltage Source Inverter) mode:
  In VSI mode, the converters operate as a voltage source. The voltage and frequency are controlled according to the demand. The droop method is used to guarantee the proper response of voltage and frequency based on the range of active and reactive power, respectively [5, 6].
  The VSI mode is often implemented in devices that control the frequency and voltage through dispatchable units, or in ESSs, according to their capacities.

It is important to mention that, when operating in islanded mode, only MCs in VSI mode can control the frequency and the voltage in the microgrid. Then, only these sources have the primary and secondary levels of control. When the microgrid is connected to the utility grid, the VSI converters change to PQ mode since the utility grid is responsible for the frequency and voltage controls.

Figure 2 shows a schematic diagram of the control in a generalized microgrid considering the presence MGCC, loads and MC in PQ and VSI mode. The dashed line represents the information flow while the continuous one points out the power flow. In the next subsections, these three levels of control are discussed.

2.1 Primary Control

The primary control is basically composed by the droop response of the converter. As stated before, the main idea is to reproduce the behavior of the traditional generation connected to the transmission system, i.e., the frequency and voltage vary when the load changes [5].

The principle of the droop method is implemented in the inverter with VSI operation mode considering the coupling P/f and Q/V (although P/V and Q/f coupling are observed in some cases in distribution systems), as shown in Eqs. (1) and (2). The coefficients m and n define the inclination of the droop, i.e. its feature [4–6].

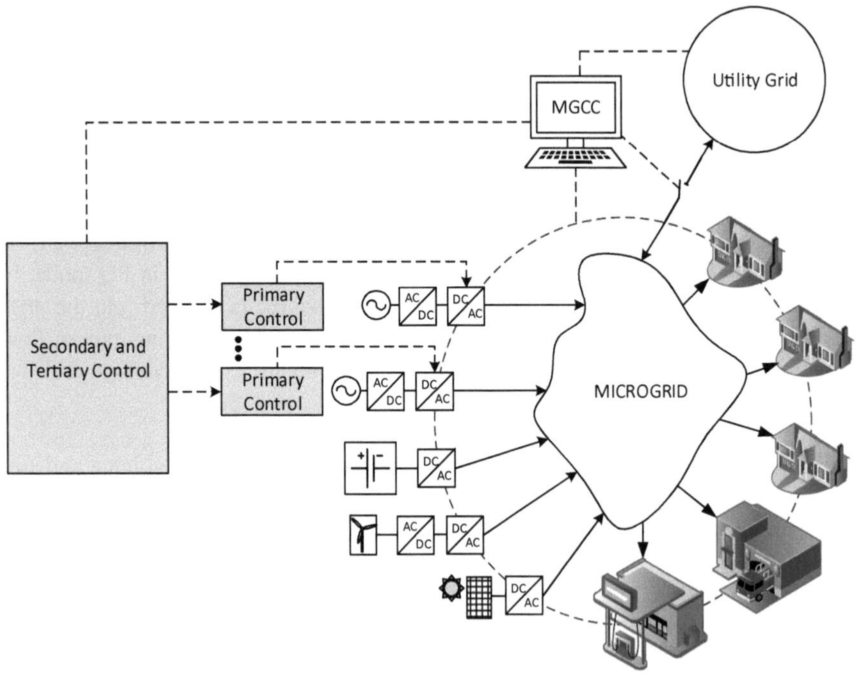

Fig. 2 Schematic diagram to a generalized microgrid

$$f = f_{\text{ref}} - m(P - P_{\text{ref}}) \tag{1}$$

$$V = V_{\text{ref}} - n(Q - Q_{\text{ref}}) \tag{2}$$

where:

f, V : the frequency and voltage outputs;
$f_{\text{ref}}, V_{\text{ref}}$: the frequency and voltage reference values;
P, Q : the demanded active and reactive power;
$P_{\text{ref}}, Q_{\text{ref}}$: the reference of active and reactive power;
m, n : the active and reactive droop coefficients.

If there is an increase in the active power demand, the frequency of the system will reduce in the proportion of the *m* coefficient; if there is a decrease in the active power demand, there is an increase in the frequency. Likewise, the voltage output in the PEC is controlled by the reactive power, according to the *n* coefficien. These responses are shown in Fig. 3.

The primary control enables the dispatchable sources to operate based on their terminal measurements to calculate the injection of active and reactive power through the droop method. This is the reason for the needless of communication in this control level, which is the main advantage of this method.

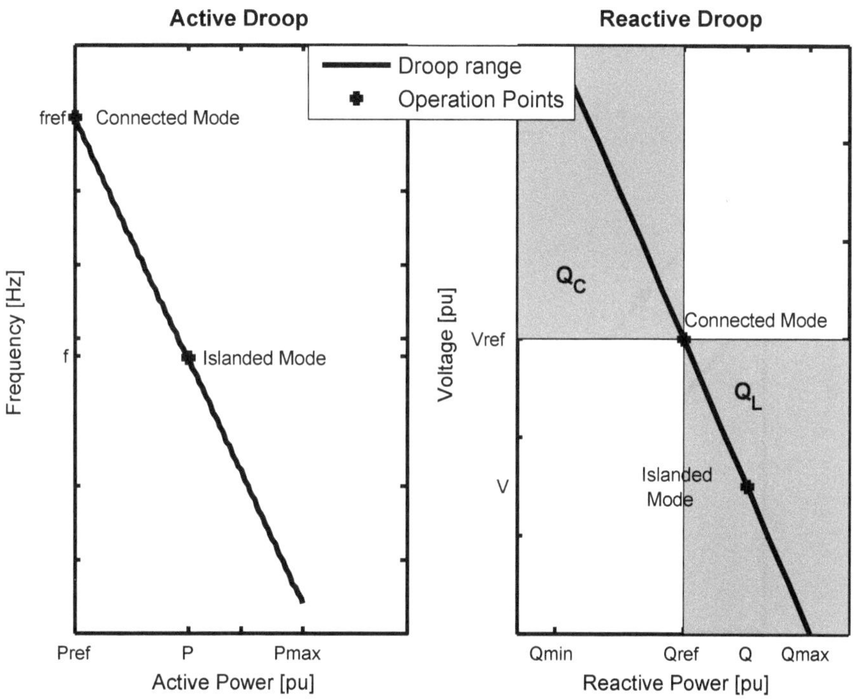

Fig. 3 Droop response

2.2 Secondary Control

The secondary control is assembled centrally at the MGCC to restore the values resulted from the droop response. In general, PI controllers are used to correct the error between the operation point and the reference [5, 7]. The communication network exchanges information between the MCs and the MGCC.

In this level of control, the operation point is moved in a vertical line. Thus, the frequency is changed while the power output is kept constant according to the demand. This is done by changing the reference of frequency and voltage from f_{ref} and V_{ref} to f^*_{ref} and V^*_{ref}. Figure 4 depicts the principle of secondary control.

More information about mathematic models, strategies of synchronism and restoration belonging the secondary control can be found in [5, 7, 8].

2.3 Tertiary Control

The tertiary control changes the reference of powers in Eqs. (1) and (2). Considering the fact that, in connected mode, the voltage and frequency are

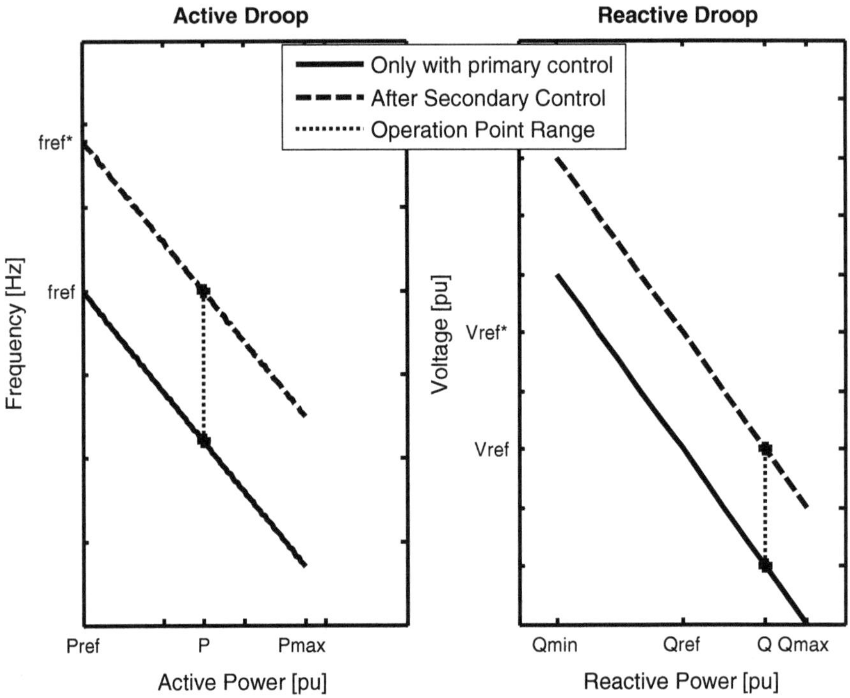

Fig. 4 Secondary control to restore the frequency and voltage

controlled by the grid, the PECs, even in VSI mode, operate similarly to PQ mode. Note that, in this case, only the frequency is kept constant during power variations because it is a global variable (the same for the whole system). This level of control also needs a communications network to receive the set points from the MGCC.

In some cases, the possibility of the MGCC does not have secondary and tertiary control should be considered. This happens mainly in rural and poor communities that do not have financial investment for implantation of a complete MGCC with a complex infrastructure of communication. Thus, to preserve the system's integrity, a simpler strategy of resources management must be implemented instead of these levels of control. A load shedding policy can also be an alternative.

A load shedding policy consists of defining a hierarchy to load shedding, i.e., non-priority loads can be shed to relieve the system's loading and improve frequency and voltage. The essential loads like hospitals can be preserved and supplied with safe and quality ranges of voltage and frequency.

3 Communications in Microgrids

As mentioned in the previous section, secondary and tertiary control communications play a key role in microgrid operation. A reliable, efficient and robust communication network must be implemented since the MGCC operation is essential for the system reliability.

The major purpose of an intelligent network is to modernize the electrical system, allowing a two-way flow of energy and information between the consumer and the utility [9].

Systems will interact with the environment and take decisions autonomously. Extensive technologies including sensors, meters and advanced control systems are installed to accomplish this autonomous operation.

Smart grids should offer communication systems that are secure, reliable and adaptive to their environment. The amount of information that will flow across the network is high and an analysis of each application is required [10].

The communication network architecture is divided into layers and there is an interaction between each layer. This interaction makes possible the inclusion of a new infrastructure model called Advanced Metering Infrastructure (AMI) [11], which will allow the bidirectional flow of information, increasing control on demand power distribution and other functionalities.

3.1 *The Design of a Communication Network for an Islanded Microgrid*

The major purpose of the microgrid network is to provide power reliably and efficiently. The smart grid has advanced communication techniques that have reliability, security, interoperability and efficiency as main features.

In this section, the main steps to designing a communication network for microgrids are described. Although it is possible to fulfill the above requirements using a plenty of technologies, the described case applies wireless communications. Wireless communications are flexible, robust, and cover large geographical areas. They are suitable for low latency applications, and also have high reliability and low infrastructure cost [12].

The basic configuration for any mobile communication system consists of three main blocks, independently of the technology used [13]. These blocks are:

- The Central and Communication Control;
- The Radio Base Station, and;
- The User.

In this model, the users are connected in a direct radio link with the radio base stations. The Radio Base Stations (RBS) are connected to the central and communications control through a physical transmission, which connects to the backhaul network, mainly by optical fibers or microwave link. The connection through the users, provided by the RBSs, is called access networks. In the investigated case, presented later, the users are the components performing monitoring and control of the grid. An access network has the purpose to provide the internet connection to the users.

There are more functions that are performed by the central control, but considering it is out of the scope, the focus will be on the functionality of the RBSs and their influence according to the coverage area and its demanded power to meet the communication demand.

3.1.1 Radio Base Stations

The basic idea of the cellular system is the division of a coverage area into smaller areas called cells. Each cell is served by a base station that through transceivers transmit power, allowing communication. In this way, it is possible to find some problems concerning the base stations, for example when transmission power increase in proportion to coverage area [14].

When the number of devices connected to the network increases, e.g. more sensors and smart meters, the power consumption of access networks also increases [15]. Cellular networks are responsible for the highest CO_2 emissions in wireless networks [16], mainly caused by Radio Base Stations (RBS) [17].

The power consumption of a base station can be modeled in two parts. The first one describes non-load dependent and the other the load dependent part [17]. The items with the greatest impact on the power consumption of a base station are: the power amplifier; the transceiver; the signal processing; the cooling system; the rectifier, and; the microwave link which connects the RBS to the backhaul network.

Each cell is divided into a number of covered sectors, which contributes to an increase in the power consumption of each RBS. Besides, a RBS is modeled by its size, and can be categorized with different models, depending on its covered area: macrocells, microcells and femtocells [18].

For the calculation of base stations, some of the parameters already mentioned above must be considered, such as environmental characteristics, coverage area, power transmission, antennas' gain, losses on receiving and transmitting and wiring losses.

In this chapter, the power consumption of all system's components for a microcell is modeled. The amplifier efficiency, n, is defined. The symbol η is the ratio between the RF (Radio Frequency) power delivered to the amplifier output and the power delivered to the amplifier input, which can be related as in Eq. (3) [17]:

$$\eta = \frac{P_{tx}}{P_{amp}}, \tag{3}$$

where P_{tx} is the amplifier output power, and P_{amp} is the input power for each antenna.

According to Eq. (3), the power consumed by the power amplifier can be determined using Eq. (4):

$$P_{amp} = \frac{P_{TX}}{\eta} \tag{4}$$

As a microcell environment is considered, the power consumption for each RBS is given by Eq. (5):

$$P_{RBS/_{micro}} = \left(P_{RBS/_{nonload}} + P_{RBS/_{load}} . L_i\right), \tag{5}$$

where $P_{RBS/_{micro}}$ is the micro RBS power consumption, $P_{RBS/_{nonload}}$ and $P_{RBS/_{load}}$ are the load and non-load power consumption, and L_i is the load factor.

The load and non-load power consumption may be assessed as Eqs. (6) and (7):

$$P_{RBS/_{nonload}} = P_{a/c} + P_{rect} \tag{6}$$

$$P_{RBS/_{load}} = P_{amp} + P_{transceiver} + P_{dsp}, \tag{7}$$

where $P_{a/c}$ and P_{rect} are the power consumption of the air conditioning and the rectifier, and P_{amp}, $P_{transceiver}$, and P_{dsp} are the power consumption of the amplifier, transceiver and digital signal processing, respectively. Notice that a microcell scheme does not support more than one sector and air conditioning is not always implemented, but as the complete power consumption model is considered, it will be included in it. Table 1 illustrates the power consumption of components of a Long-Term Evolution (LTE) RBS.

The study presented in the next section, LTE is the adopted technology [18]. LTE is a wireless broadband technology, also known as the fourth generation

Table 1 Power consumption parameter for a LTE base station [17]

Component	Parameters	Power consumption
Digital signal processing	P_{dsp}	100 W
Efficiency	η	12%
Input power	P_{TX}	33 [dBm]
Power amplifier	P_{amp}	156 W
Transceiver	$P_{transceiver}$	100 W
Rectifier	P_{rect}	100 W
Air conditioning	$P_{a/c}$	60 W
Load factor	L_i	1

mobile (4G). Developed by 3GPP (Third Generation Partnership Project), this is one of the recent standardization of mobile telephony. The main features of LTE technology are the transmission of data at high transmission rate and low latency.

3.1.2 Planning a Mobile Network for a Microgrid

The network planning process is carried out to extract the highest performance coverage with the least amount of equipment possible. Several steps have to be fulfilled to achieve these goals. The first planning stage is to obtain information such as the desired coverage area, the distribution of RBSs in the coverage area, and the required Quality of Service (QoS).

The amount of users who will access the communication networks using RBS should be assessed. This represents 60–80% of energy consumption in a radio access network [19]. The result will give a minimum density of RBS to be installed to meet the desired coverage area. The following topics will describe step by step the cell planning of a micro network for a microgrid considering the LTE as a candidate technology.

The first stage is to calculate the Link Budget, where the maximum propagation loss allowed for users who may receive the signal at the edge of this system is calculated.

After designing the Link Budget, the next step is to determine the radius of the cell, allowing an estimation of the number of base stations required to cover the desired area to be connected. Finally, it is possible to assess the total power consumption of communications network.

A detailed analysis of the site where the LTE wireless communication system will be implemented is necessary, taking into account characteristics of the environment, the place where the RBSs will be installed, topography, and population density.

During the designing of cellular networks, it is necessary to calculate both the downlink and the uplink link budget [20]. The uplink is important because it measures the power and sensitivity of the user, while the downlink considers coverage constraints. As the main objective here is to provide communications for the maximum area available, the worst value of sensitivity is considered.

After determining the coverage area, the next step is assessing the link budget for both the downlink and the uplink. Then, it is possible to determine the maximum loss per way, which is an important variable to calculate the radius of the cell [20]. Equation (8) assesses the link budget in decibels [dB]:

$$L = P_{\mathrm{tx}} + G_{\mathrm{tx}} - L_{\mathrm{tx}} - SNR_{\mathrm{required}} - S_{\mathrm{rx}} + G_{\mathrm{rx}} - L_{\mathrm{rx}} + G_{\mathrm{dv}} \tag{8}$$

where, L is the path loss for downlink, P_{tx} is the transmission power [dBm], G_{tx} is the transmit antenna gain [dBi],L_{tx} are the loss of transmission [dB], SNR_{required} is the Required Signal Noise Ratio [dB], S_{rx} is the required receiver sensitivity [dB],

G_{rx} is the receiver antenna gain [dBi], L_{rx} are the losses in reception [dB], G_{dv} is the diversity gain [dBi].

Therefore, the calculation of the maximum propagation loss for the link propagation is described in Eq. (9):

$$MP_{\text{propagation}} = L - L_{\text{penetration}} - BL - F[dB](9) \quad (9)$$

where $L_{\text{penetration}}$ indicates the loss of radio signal due to obstruction, BL is the losses due to obstruction by a body and F is the fade margin, it means the minimum acceptable receiver level. The maximum path loss calculation uses Table 2 values [23].

3.1.3 Range and Number of Base Stations

For the design of wireless communication systems, it is necessary to choose a proper propagation model. These models aim to provide estimates of propagation loss considering the distance between transmitter and receiver, terrain factors, the height of the transmitting and receiving antennas and the frequencies used.

The basic SUI (Stanford University Interim) model proposed by the IEEE is applied. It is based on Erceg's model [21] and the range is assessed using Eq. (10) [23]:

Table 2 Parameters for path loss calculation

LTE link budget and path loss parameters	
Frequency	2600 MHz
Maximum input power of base station	33 [dBm]
RBS's antenna gain	4 [dBi]
RBS's feeder loss	2 [dB]
Fading margin	8.1 [dB]
Bandwith	20 MHz
Penetration loss	2.6 GHz 20 [dB]
Thermal noise	−174 [dBm]
SNR sensitivity	−92 [dBm]
Installation loss	2 [dB]
Duplexing	TDD
Path loss model	SUI—Erceg
Coverage requirement	98%
Heigth of base station	30 m
Heigth of mobile station	2 m
Area type	Urban
Body loss	4 [dB]

$$R = f^{-1}(L - F|f, h_{\text{user}}, h_{\text{BS}}), \tag{10}$$

where R is the maximum range in meter, h_{user} and h_{BS} are the antenna and RBS height in meters. After this stage, it is possible to determine the number of RBS for a predetermined circular area applying Eq. (11) [23]:

$$\text{Number RBS} = \frac{\text{Desired Coverage Area}}{\pi * R^2} \tag{11}$$

So, the maximum power consumption from all RBSs (P_{max}) is assessed according to Eq. (12):

$$P_{\text{max}} = \sum_{i=1}^{\text{NumberBS}} P_{\text{RBS/micro}} \tag{12}$$

where $\boldsymbol{P}_{\text{RBS/micro}}$ is the power consumption from each RBS and **NumberBS** is the total number of RBS.

4 Operation of Islanded Migrogrids: A Case Study

In this section, an example of a microgrid operating in islanded mode is presented, considering the previous topics (except the secondary and tertiary controls). The system used here is the IEEE 13 Bus Test Feeder. However, some modifications are made to the original version, as listed below.

- The equivalent power in each bus is the average of all phases in this bus;
- Spot loads are equally divided between the two buses in the corresponding line;
- The line impedance is represented only by the positive sequence;
- Generation units and Energy Storage System are inserted in this distributed system to characterize a microgrid.

The intention of these simplifications is to transform the original unbalanced system into a balanced one with some characteristics of a microgrid. This enables one to use the method proposed in [6] to find de voltages, angles and frequency deviation. The complete system's data used here can be found in [24]. Note, however, that these assumptions are considered for simplicity, since [25] applies the droop method for three-phase unbalanced microgrid.

For the design and implementation of a homogeneous cellular network in the microgrid, it must be considered the steps which were described in the previous section. As the power amplifier efficiency is given the input power, it can be derived from Eqs. (3) and (4) as described in Eq. (13),

$$P_{\text{amp}} = \frac{(10^{33/10}) * 0.001}{0.12} = 16.62 \text{ W} \tag{13}$$

From Table 1 and Eqs. (6) and (7), the load and non-load power consumptions are assessed in Eqs. (14) and (15):

$$P_{\text{RBS}/_{\text{nonload}}} = 60 + 100 = 160\,\text{W} \tag{14}$$

$$P_{\text{RBS}/_{\text{load}}} = 16,62 + 100 + 100 = 216.62\,\text{W} \tag{15}$$

From Eq. (5) and the results accomplished in Eqs. (14) and (15) for a maximum load factor, the power consumption of one microcell base station is assessed according to Eq. (16),

$$P_{\text{RBS}/_{\text{micro}}} = (160 + 356.1) = 376.62\text{W} \tag{16}$$

So, the total power consumption from one RBS is 376.62 W. As the proposed is to determine the number of RBS to cover the microgrid system, Now, it is time to calculate the link budget for both downlink and uplink.

Considering Table 2 and from Eq. (8), the path loss for the downlink is given by Eq. (17),

$$L_{\text{down}} = 33 + 4 - 2 - 0 - (-92) + 0 - 0 + 0 = 127\ [\text{dB}] \tag{17}$$

From Eq. (9), it is possible to calculate the maximum propagation loss which will be taken into account the number of necessary base stations to achieve real-time cellular communications for the case of study, according to Eq. (18):

$$MP_{\text{down}} = 127 - 20 - 4 - 8.1 - 94.9\ [\text{dB}] \tag{18}$$

As the main objective here is to provide communications for the maximum area available, the worst value of sensitivity is considered, it is calculated the link budget for the uplink and it is chosen the worst scenario for further considerations, the same steps are done and the link budget assessed for the uplink is shown in Eqs. (19) and (20),

$$L_{\text{up}} = 139.5\ [\text{dB}] \tag{19}$$

And

$$MP_{\text{up}} = 104.9\ [\text{dB}] \tag{20}$$

As it can be seen above the worst scenario is the in the uplink, as the users must be connected into the edge of the cell. The next step is to evaluate the maximum range of a RBS given by Eq. (10) and for data retrieved from Table 2, it can be assessed as in Eq. (21)

$$R = 10^{(\mathrm{L-M}\,|f, h_{\mathrm{user}}, h_{\mathrm{BS}})} = 0.2543\ \mathrm{m} \tag{21}$$

The desired coverage area is illustrated in Fig. 5 below,

From the Fig. 5 and Eq. (11), it is determined the number of base stations for the desired area as in Eq. (22),

$$\mathrm{NumberRBS} = \frac{0.6}{\pi * (0.25)^2} = 3 \tag{22}$$

And finally, the total power consumption for a LTE system network implemented for communications in a microgrid, is assessed in Eq. (23), as described earlier in Eq. (12),

$$P_{\max} = 3 * 376.62 = 3 * 376.62 = 1130\ \mathrm{W} \tag{23}$$

After discussing the communications network design for this specific study case, the modifications to the power grid need to be described further. Three dispatchable units are considered, one ESS, three renewable units and three base stations. The topology of the proposed microgrid is illustrated in Fig. 6.

The RBSs were dimensioned considering the power demand and the size of the system. According to the approach described in the previous section and to cover

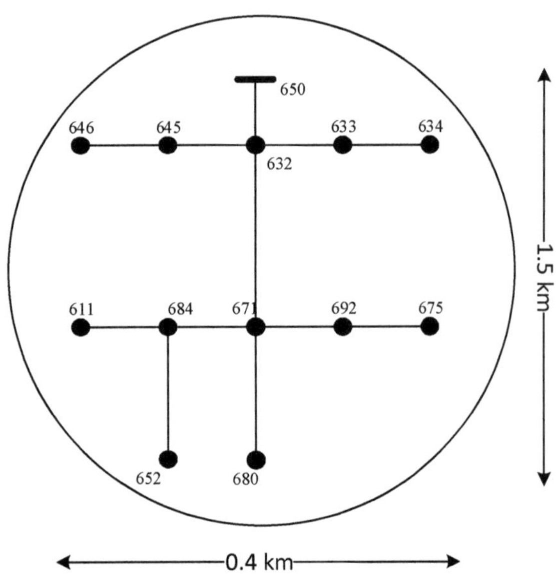

Fig. 5 Modified IEEE 13 bus system coverage area

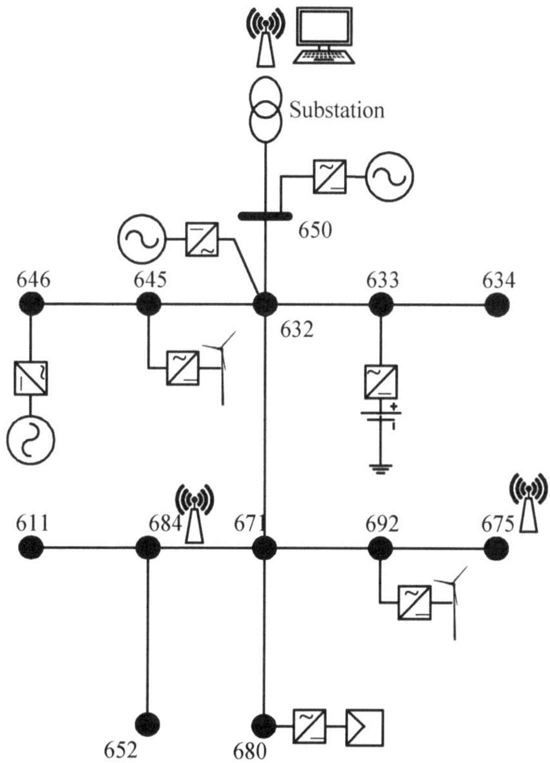

Fig. 6 The Microgrid test system

the whole area, three base stations should be installed. Each one consumes 1.13 kW and they are located as depicted in Fig. 6.

During the islanded operation, the power supply to all loads must be kept. However, a large piece of the power generation from the microgrid is based on small-scale renewables. Their intermittent feature may cause a large variation in the power generation, which is regulated by the presence of ESSs.

Connected to the microgrid, there are high priority customers which must be supplied at any time. Therefore, an operation framework must be developed to ensure the supply of the highest priority loads by applying some actions, e.g. load shedding schemes.

Some important pieces of information about loads values and generation schedule, both dispatchable and non-dispatchable, are presented in Tables 3 and 4, respectively. In Table 4, the droop coefficients and the maximum power of the dispatchable units are also presented.

Photovoltaic arrays and wind turbines are considered as non-dispatchable units. In this case, they operate in PQ mode. The models used for photovoltaic arrays and wind turbines are presented in [26, 27]. The ESSs have PQ inverters that control the injected power based on the system's demand. The data regarding renewable sources and ESS are summarized in Table 5.

Table 3 Load characteristics

Bus	Equivalent demand [kW]	Power factor	Priority
650	10	0.92	High
632	34	0.87	Low
633	0	0.92	Low
634	134	0.83	Low
645	57	0.81	Medium
646	77	0.87	Low
671	400	0.85	Low
684	0	0.92	High
611	57	0.9	Low
652	43	0.83	Low
680	0	0.92	Low
692	57	0.75	Medium
675	281	0.87	High

Table 4 Dispatchable units generation schedule

Bus	Droop coefficient *m*	Droop coefficient *n*	Maximum power [pu]
650	Variable	Variable	0.6
632	0.025	0.00001	0.6
633	0.025	0.00001	0.6

Table 5 Renewable and ESS configuration

Solar				
Bus	n_s	n_p	P[pu]	
680	20	20	0,35	
Wind turbine				
Bus	ω_{ci} [m/s]	ω_{co} [m/s]	ω_r [m/s]	P[pu]
645	4	12	8	0,1
692	4	12	8	0,1
Batteries				
Bus	$Cap_{m\acute{a}x}[pu.hr]$	$\mathrm{Cap}_{\min}$ [pu.hr]	P[pu]	
633	0.03	0.002	0,01	

In Table 5, n_s and n_p are the numbers of serie and parallel photovoltaic arrays. The cut-in, cut-off and rated speed of the wind turbine are represented by ω_{ci}, ω_{co} and ω_r, respectively. Besides, the injected power by photovoltaic arrays depends on cell temperature and solar irradiance; the injected power by wind turbine depends on the wind speed.

In the example presented here, the operation of an islanded microgrid is investigated during one random day. The weather conditions and the load profile are illustrated in Fig. 7.

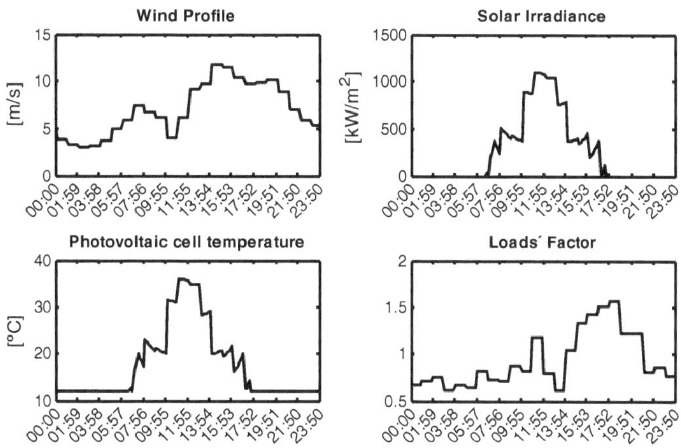

Fig. 7 Weather conditions and load profile

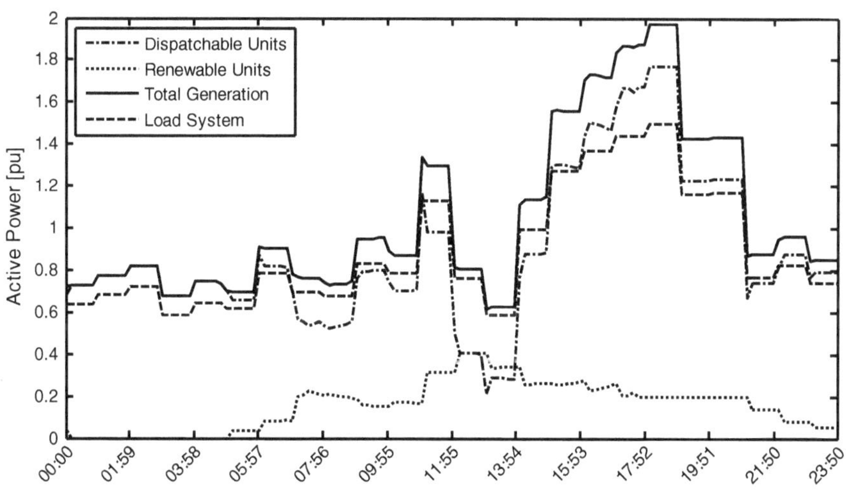

Fig. 8 The power flow result for the microgrid

Figure 8 shows the time continuous power balance result for the microgrid. Note that, the total generation is a little higher than the load, since the total generation must supply the losses. The dispatchable generation is the responsible for the power balance of the system, however, the renewable generation has a considerable participation. Whenever there is an increase in renewable penetration, the dispatchable generation is reduced. On the other hand, dispatchable generation increases the amount of generation during certain periods where the renewable generation is scarce.

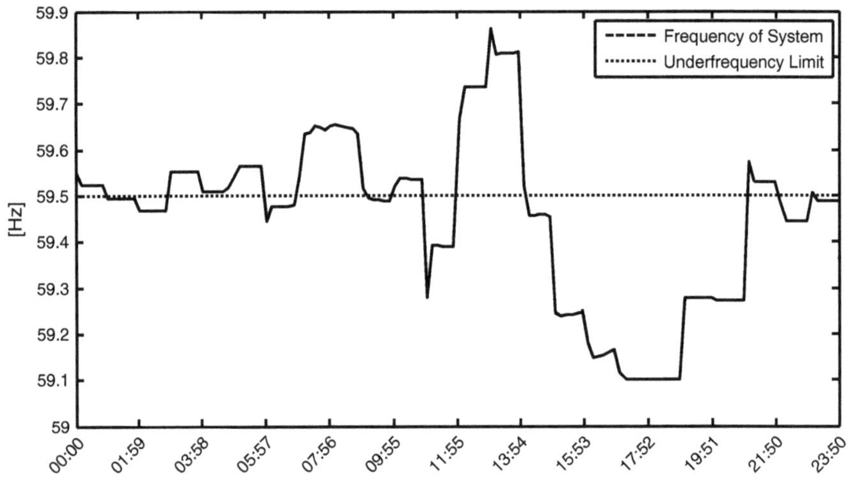

Fig. 9 Frequency profile of the microgrid

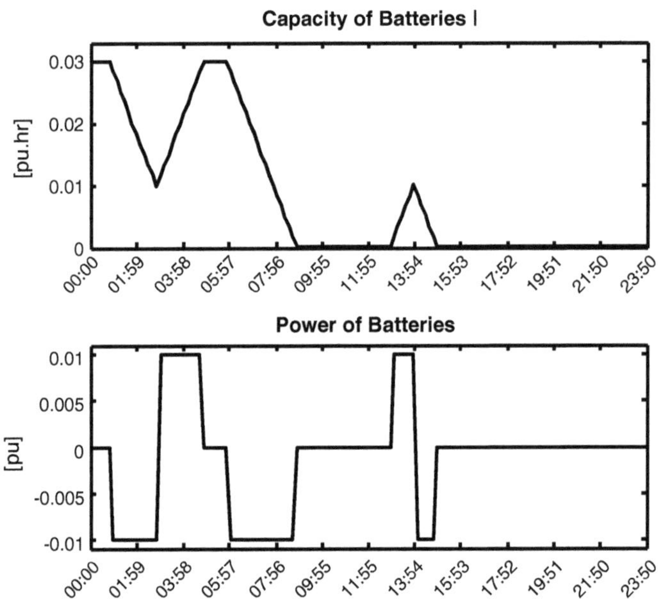

Fig. 10 The behavior of the ESS

Figure 9 shows the frequency response. The under-frequency limit is 59.5 Hz, while the over-frequency limit is 60.5 Hz, considering that the nominal frequency of the system is 60 Hz. When the frequency is over the limit, there is an excess of generation. In this scenario, the batteries are connected as load, consuming power to charge to its maximum capacity. On the other hand, when the frequency is under

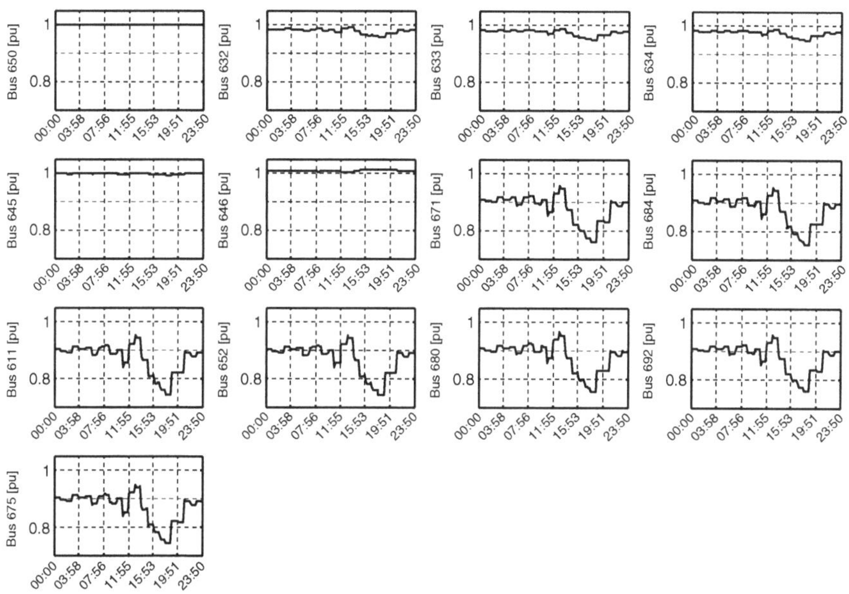

Fig. 11 Voltage profiles for all buses

the limit, the batteries are injecting power into the system and discharging until the minimum capacity is reached. This behavior is depicted in Fig. 10.

Finally, Fig. 11 shows the voltage profiles of all buses. It is possible to see that Buses 632 and 646% a variable voltage profile during the day even being generation buses, i.e., PV buses in the power flow. This occurs because of the reactive droop response, that change the voltages based on the reactive power demand. While the voltage at the reference bus is kept constant, all others buses are PQ and their voltage changes according to the load demand and power generation during the day.

5 Conclusions

This chapter presents some background on the operation of an islanded microgrid. Considering a centralized control approach, the primary, secondary, and tertiary control levels are described. The communications requirements for a microgrid are also presented and the designing of a communication network for a microgrid is described step-by-step.

Although there is a plenty of technologies that may be applied to develop a communication network, the wireless technology, in particular the LTE, is recommended . This recommendation is supported by the infrastructure costs and other technical requirements, e.g. coverage area, quality of service, and reliability.

The Operation of an islanded microgrid during one day is investigated in a case study. The LTE communication network for this case is also designed according to the previously described steps. The power consumption of the RBSs are a concern for this scenario, since even the communication network is important, its power consumption affects the microgrid energy autonomy in this emergency scenario.

References

1. H. Farhangi, "The Path of the Smart Grid," *IEEE Power & Energy Magazine*, vol. 8, no. 1, pp. 18–28, 2010.
2. A. Khodaei "Resiliency-Oriented Microgrid Optimal Scheduling," *IEEE Transactions on Smart Grid*, vol. 5, no. 4, pp. 1584–1591, 2014.
3. B. Starfield, H. Lotfi, A. Khodaei, and S. Bahramirad, "State of the Art in Research on Microgrids: A Review," *IEEE Access*, vol. 3, no. 1, pp. 890–925, 2015.
4. J. A. P. Lopes, C. L. Moreira, and A. G. Madureira, "Defining Control Strategies for MicroGrids Islanded Operation," *IEEE Transactions Power Systems.*, vol. 21, no. 2, pp. 916–924, 2006.
5. J. M. Guerrero, J. C. Vasquez, J. Matas;, L. G. de Vicuña, and M. Castilla, "Hierarchical Control of Droop-Controlled AC and DC Microgrids — A General Approach Toward Standardization," *IEEE Transactions. Industrial Electronics*, vol. 58, no. 1, pp. 158–172, 2011.
6. A. C. Zambroni de Souza, M. Santos, M. Castilla, J. Miret, L. G. De Vicuña, and D. Marujo, "Voltage security in AC microgrids : a power flow-based approach considering droop-controlled inverters," *IET Renewable Power Generation*, pp. 1–7, 2015.
7. Y. Jia, D. Liu, and J. Liu, "A Novel Seamless Transfer Method for a Microgrid Based on Droop Characteristic Adjustment," in *IEEE Internacional Power Electronics and Motion Control Conference*, 2012, pp. 362–367.
8. C. Jin, M. Gao, X. Lv, and M. Chen, "A Seamless Transfer Strategy of Islanded and Grid-connected Mode Switching for Microgrid based on Droop Control," in *IEEE Energy Conversion Congress and Exposition*, 2012, pp. 969–973.
9. V. C. Gungor, D. Sahin, T. Kocak, S. Ergut, C. Buccella, C. Cecati, and G. P. Hancke, "A Survey on smart grid potential applications and communication requirements," *IEEE Transactions Industrial Informatics*, vol. 9, no. 1, pp. 28–42, 2013.
10. K. Tomsovic, D. E. Bakken, V. Venkatasubramanian, and A. Bose, "Designing the Next Generation of Real-Time Control, Communication, and Computations for Large Power Systems," *Proceedings of IEEE*, vol. 93, no. 5, pp. 965–979, 2005.
11. R. R. Mohassel, A. Fung, F. Mohammadi, and K. Raahemifar, "A survey on Advanced Metering Infrastructure," *International Journal of Electrical Power Energy Systems*, vol. 63, pp. 473–484, 2014.
12. V. Aravinthan, B. Karimi, V. Namboodiri, and W. Jewell, "Wireless communication for smart grid applications at distribution level—Feasibility and requirements," *IEEE Power Energy Society General Meeting 2011 IEEE*, pp. 1–8, 2011.
13. Michel Daoud Yacoub. 1993. *Foundations of Mobile Radio Engineering* (1st ed.). CRC Press, Inc., Boca Raton, FL, USA.
14. M. Gruber, O. Blume, D. Ferling, D. Zeller, M. A. Imran, and E. C. Strinati, "EARTH — Energy Aware Radio and Network Technologies," *2009 IEEE 20th International Symposium on Personal Indoor Mobile Radio Communications*, January, pp. 1–5, 2009.
15. S. N. Roy, "Energy Logic : A Road Map to Reducing Energy Consumption in Telecommunications Networks," IEEE 30th International Telecommunications Energy Conference,, pp. 2–10, 2008.

16. A. Fehske, J. Malmodin, G. Biczók, and G. Fettweis, "The Global Footprint of Mobile Communications–The Ecological and Economic Perspective," *IEEE Commununications Magazine*, August, pp. 55–62, 2011.
17. M. Deruyck, W. Joseph, and L. Martens, "Power consumption model for macrocell and microcell base stations," Transactions on Emerging Telecommunications Technologies, August 2012, pp. 320–333, 2014.
18. ETSI, "TR 36.921 - V9.0.0 - LTE; Evolved Universal Terrestrial Radio Access (E-UTRA); FDD Home eNode B (HeNB) Radio Frequency (RF) requirements analysis (3GPP TR 36.921 version 9.0.0 Release 9)," *Eur. Telecommun.*, vol. 0, pp. 0–44, 2010.
19. G. Koutitas and P. Demestichas, "A Review of Energy Efficiency in Telecommunication Networks," *Telfor Journal*, vol. 2, no. 1, pp. 2–7, 2010.
20. Theodore Rappaport. 2001. *Wireless Communications: Principles and Practice* (2nd ed.). Prentice Hall PTR, Upper Saddle River, NJ, USA.
21. V. Erceg et al., "*Channel Models for Fixed Wireless Applications*" IEEE802.16.3c-01/29r4, Broadband Wireless Working Group, IEEE P802.16, 2001.
22. TSGR. TS 136 104 - V11.4.0 - LTE; Evolved Universal Terrestrial Radio Access (E-UTRA); Base Station (BS) radio transmission and reception (3GPP TS 36.104 version 11.4.0 Release 11), 2013.
23. M. Deruyck, W. Vereecken, E. Tanghe, W. Joseph, M. Pickavet, L. Martens, and P. Demeester, "Power consumption in wireless access networks," *European Wireless. Conference*, pp. 924–931, 2010.
24. IEEE Distribution Feeders Analysis Subcommittee. Distribution Test Feeders. Available on: http://ewh.ieee.org/soc/pes/dsacom/testfeeders/index.html.
25. M. M. A. Abdelaziz, H. E. Z. Farag, E. El-Saadany, Y. A.-R. I. Mohamed. "A Novel and Generalized Three-Phase Power Flow Algorithm for Islanded Microgrids Using a Newton Trust Region Method". IEEE Transactions on Power Systems, V. 1, No. 28, pp. 198–201, 2013.
26. M. Katsanevakis. "Modelling the photovoltaic module", 2011 IEEE International Symposium on Industrial Electronics (ISIE), June de 2011, pp. 1414–1419.
27. E. Vittal; M. O'Malley; A. Keane; "A Steady-State Voltage Stability Analysis of Power Systems With High Penetrations of Wind". IEEE Transactions on Power Systems, vol. 25, pp. 433–442, 2010.

Islanding of Energy System

Jayesh Joglekar

Abstract The availability of electrical energy ensures development and progress of a country while its non-availability means economic stagnation or 'backtracking'. The condition of 'No Power' is the costliest state and leads to social, economical and production loss. Power system blackout means inconvenience and hardship to society. Therefore, reliable and disturbance-free electric power supply is recognised as a key to societal progress throughout the world. Uninterrupted power supply is therefore essential for a higher national productivity and improvement in the social structure. Grid failure can occur due to a variety of causes such as: continued over-stressing, low frequency operation, increased load demand, peaking characteristics of the demand, ineffective load control, reduced level of security due to opening of interstate lines and violation of grid discipline. Separating healthy part in the system during fault at pre-determined points is called as islanding. This is to ensure supply of power to industries to reduce economic losses. In a complex system, intentional islanding becomes necessary to save healthy part to recover from the blackout. In this research work, Thermal Power Stations (TPS) are considered for intentional islanding as they have the highest installed capacity in India. Due to the effect of grid disturbance, a generating station usually experiences a shortfall of power for running its auxiliaries.

Keyword Islanding · Restoration · Bottom-up approach

1 Introduction

Large interconnected power systems may be seriously affected by severe occurrences that could lead to a cascade of automatic actions. The power systems operated by the utilities in developing countries suffer from a large gap between demand and generation, inadequate transmission capacity and non-uniform location of the load

J. Joglekar (✉)
Maharashtra Institute of Technology, Pune, India
e-mail: joglekar@ieee.org

B. Azzopardi (ed.), *Sustainable Development in Energy Systems*,
DOI 10.1007/978-3-319-54808-1_11

centers and generating stations. In most of the cases, occurrences of faults in such systems end up with the worst consequence, which is complete blackout. After occurrence of severe system disturbances, the system may split into parts, which may or may not survive depending on the load generation balance. The part system containing the generation sources and certain loads, which are planned to be separated from the main grid during system disturbance at preconceived points either through under frequency and/or directional power relays are called 'Islands'. The availability of electrical energy ensures development and progress of a country while its non-availability means economic stagnation. The condition of 'No Power' is the costliest state than no power. Uninterrupted power supply is therefore essential for a higher national productivity and improvement in social structure. The power grid failures, causing widespread blackouts, can occur due to a variety of causes such as: continued over-stressing, low frequency operation, increased load demand, peaking characteristics of the demand, ineffective load control, reduced level of security due to opening of interstate lines and violation of grid discipline.

2 Islanding Methods

The concept of islanding in power system is referred as supplying electricity to the load isolated from grid using the generator within the system. The system stability during widespread cascading grid failure can be restricted using islanding of healthy part within the system to recover the restoration. Different approaches and logical methodologies of power system restoration have common requirement of start-up power or black start power. There are broadly two methods for islanding as per the situation, namely planned or intentional islanding and forced or un-intentional islanding. The idea of intentional islanding is concurrent to all ways of quick grid restoration. Intentional islanding is joint venture of sufficient generation, stable load and a strong communication system. The generation in islanded sub-system is mostly unconventional and hence its modular structure can be located near the load for better performance. This chapter discusses a new algorithm and its comparison with conventional methodologies to find out the location for modular source. The forced islanding is referred as a corrective measure to save small part of grid or system to avoid cascade tripping or complete blackout. This method is operated for small time duration depending on system condition. Due to emergency islanding, in this method power number (MW/Hz) is very small, hence this method is very unstable in practice.

3 Islanding Conditions: Proposed New Algorithm

The proposed algorithm is divided into two parts. In first part, contribution of the source in MW is calculated for each node and based on that, the node ranking is done. In this ranking, the node with maximum contribution of source is given first rank and given the first preference for intentional islanding.

The second part of the algorithm is executed after obtaining real time node and source weight (power in MW). With the help of computer program, nodes are short-listed using node ranking. The node ranking is based on the ratio of source weight to node weight which lies between 1.0 and 1.1. The total ranking for short-listed nodes decides boundary of intentional islanding scheme.

The generalized formula is:-

$$1.0 < \frac{\sum_{i=1}^{i=\infty} G_i}{\sum_{k=1}^{k=\infty} L_k} \leq 1.1 \tag{1}$$

where,

G_i value of ith generation available in mw
L_k value of kth radial feeder load in MW
I number of generators in the system
K number of radial feeders in the system

Steps

1. Read network as a directional graph model
2. Read source and node weight
3. Consider one source at a time and its weight named as 'a'

Part-I

4. Run the DC load flow for the given network
5. Redraw the directional rooted graph model with 'a'
6. Draw sub-graph by considering permutation—combination of graph branches connected with node
7. Calculate ratio of average incoming weight to each node weight in each sub graph
8. Calculate average ratio for each node
9. Average ratio of each node is source 'a' contribution in each node

End
Part-II

10. Read source contribution in each node
11. Assign rank to each node based on source contribution
12. Allot first rank to the node with maximum source contribution and subsequently allot ranking to the other nodes.
13. Calculate ratio in limit condition as $\left[1.0 \leq \frac{\text{Source weight}}{\text{Node weight}} \leq 1.1\right]$ and make a group of such nodes
14. Choose the highest ranking group of nodes for intentional islanding

End
The flow chart of algorithm is shown in Fig. 1.

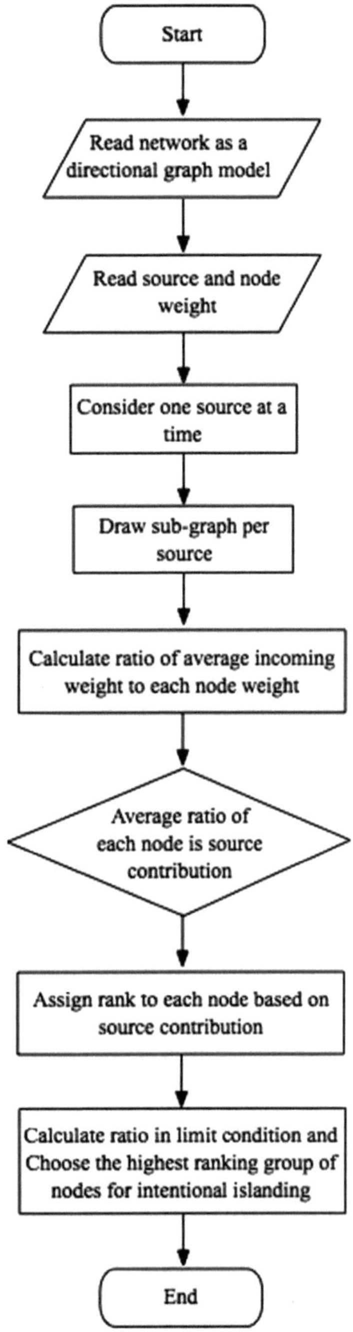

Fig. 1 Flow chart of algorithm

4 Intentional Islanding: Demonstrating the New Algorithm

Figure 2 represents single line diagram of hypothetical system for understanding the algorithm. In the network, generator is connected with seven different loads (L_1–L_7). One generator is considered at a time for this algorithm graph theory line diagram as shown in Fig. 3. Figure 4 shows load contribution of a generator in each load. The load ranking is assigned based on contribution. Rank 1 is given to the load which has maximum generator contribution. In the same way, all loads connected in the

Fig. 2 Single line diagram of a hypothetical system

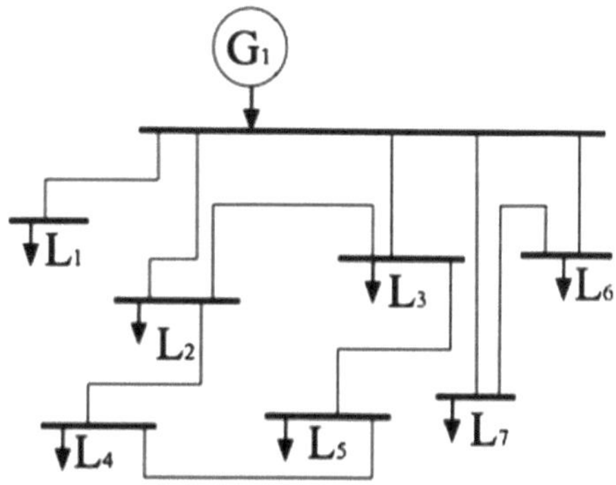

Fig. 3 Generator and load connection

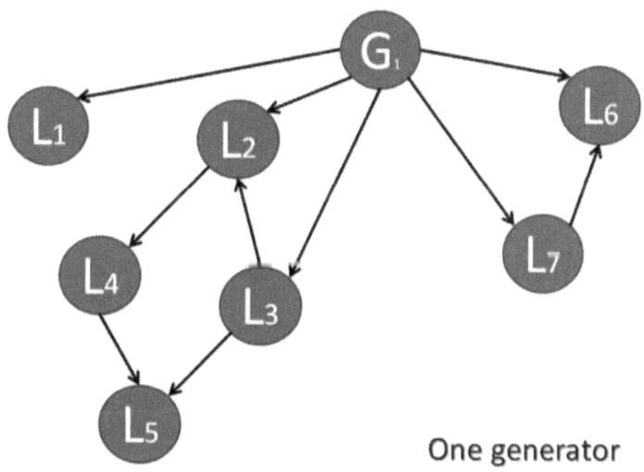

Fig. 4 Generator and load pair ranking

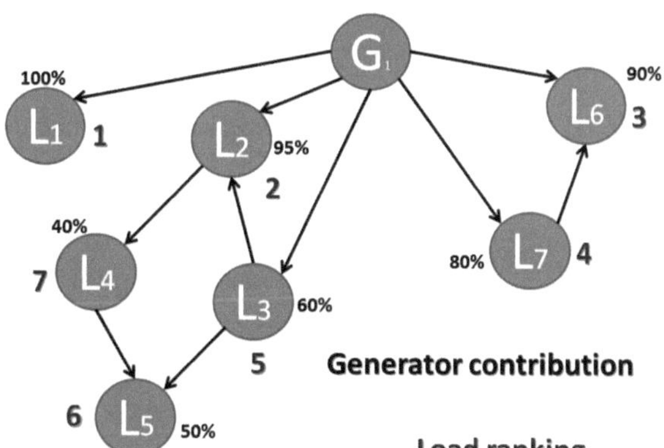

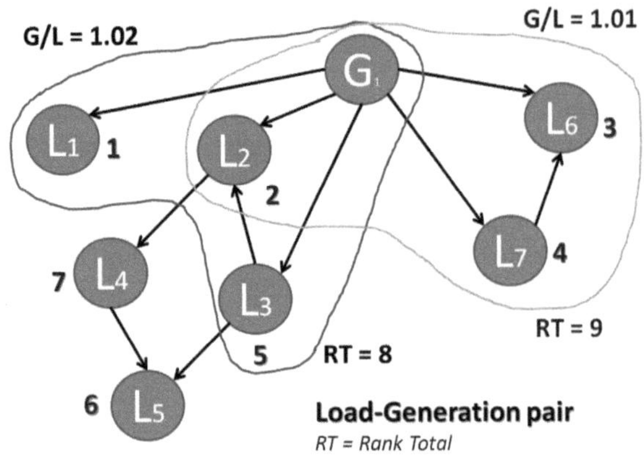

Fig. 5 Development of Generator and load pair

network are given ranking. Figure 5 shows the generator and load combinations which satisfy the G/L ratio as mentioned above. The rank total (RT) is calculated based on ranking. The generator—load combination with lowest RT is selected for intentional islanding and active source installation as it has load with maximum generator contribution. In this way location of active source is decided.

5 Application of Graph Theory in Islanding

The optimal selection of group of loads near to a generator in which it has maximum contribution is necessary for stable islanding procedure. Two methods are used for load selection. In the first method load selection is done which will satisfy Eq. 1. This method helps to short-list loads from connected network for a particular generator. The second method is for selection of load (s) in which particular generator has maximum contribution. For this method, graph theory is used [1]. The procedure for applying graph theory is as explained below:

1. Run DC Load Flow for network with generator and load.
2. Convert the network into graph: Generator bus will be solid dots, load bus will be hollow dots, transmission line will be branches of graph and direction of power flow will be indicated by directional arrow on every branch.
3. Next step is to separate the graph into directed sub-graphs with generation as its roots.
4. Find generator and load combinations which satisfy ratio of generator power to load power in between 1.0 and 1.1. (please refer Eq. 1).
5. Find out contribution of a particular generator by formula: C_{ij} = Receiving power/Load power for all sub-graphs, where i is the generator and j is the load.
6. Averaging all the sub-graph values, contribution of all generators in the network for a particular load can be determined. The above procedure is applied to IEEE 14 bus test system as shown in Fig. 6.

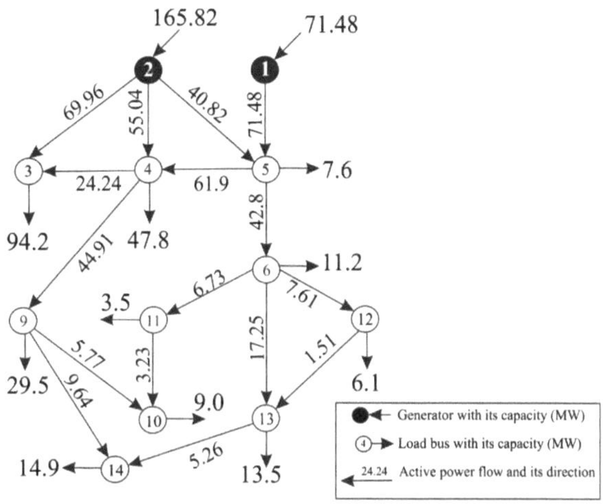

Fig. 6 Graph Theory model for IEEE 14 bus system

Table 1 Generator contribution in IEEE 14 bus system

Bus no.	Load (MW)	Generator contribution			
		G1	%age	G2	%age
2	21.7	14.08	64.88	7.62	35.12
3	94.2	77.85	82.64	16.35	17.36
4	47.8	41.86	87.57	5.94	12.43
5	7.6	7.04	92.63	0.56	7.37
6	11.2	10.45	93.30	0.75	6.70
9	29.5	25.81	87.49	3.69	12.51
10	9.0	8.09	89.89	0.91	10.11
11	3.5	3.26	93.14	0.24	6.86
12	6.1	5.68	93.11	0.42	6.89
13	13.5	12.59	93.26	0.91	6.74
14	14.9	13.36	89.66	1.54	10.34

A sample calculation using above method is shown below for load bus 3:

1. Load bus: 3
2. Generator bus: 1
3. C_{ij} = Receiving power/Load power = 69.96/94.20 = 0.7426. Generator 1 contribution in load bus 3 is 74.26%.
4. Generator bus: 2
5. C_{ij} = Receiving power/Load power = 24.24/94.20 = 0.2573. Generator 2 contribution in load bus 3 is 25.73%.

The generator contribution for each load in IEEE 14 bus system is calculated and summarized in Table 1. From the tabular results it is clear that for every load in the network one particular generator has maximum contribution. The loads with maximum contribution of generator are paired and such load-generator pairs are considered for sub-system connection and islanding.

The above approach is applied for IEEE 14 bus system. Initially in Case-I, only one generator G1 is considered with the load connected as shown in Fig. 7 and corresponding loads in Table 2.

Using computer program, combination of loads with generator are obtained which satisfies Eq. 1. The load combinations having G/L ratio near to 1.0 and near to 1.1 are tabulated in Table 3.

Similarly considering generator G2 as case-II, for finding out load connected as shown in Fig. 8 and corresponding loads in Table 4.

The load combinations having G/L ratio near to 1.0 and near to 1.1 are shown in Table 5.

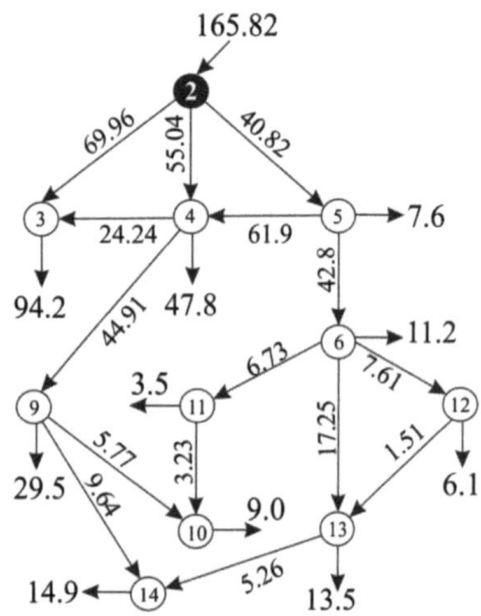

Fig. 7 Generator and load pair in IEEE 14 bus system with Generator 1

Table 2 Generator-1 and load pair in IEEE 14 bus system

Generator: 1	Load: 10	
G [1]: 165.82	L[1]: 94.19	L[6]: 9.0
	L[2]: 47.80	L[7]: 3.50
	L[3]: 7.60	L[8]: 6.10
	L[4]: 11.20	L[9]: 13.50
	L[5]: 29.50	L[10]: 14.90

Table 3 Generator-1 and load pair in IEEE 14 bus system satisfying Eq. 1

Sr. no	Combination		
	G/L ratio	Generator	Load (s)
1	1.098873	G1	L1 + L3 + L5 + L8 + L9
2	1.000121		L1 + L3 + L5 + L8 + L9 + L10

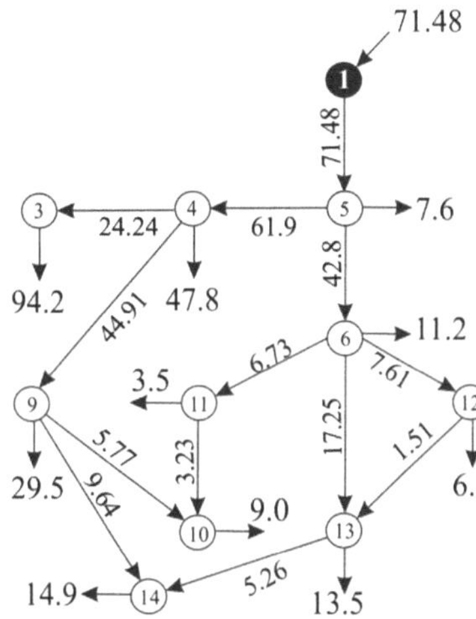

Fig. 8 Generator and load pair in IEEE 14 bus system with Generator 2

Table 4 Generator-2 and load pair in IEEE 14 bus system

Generators: 1	Load: 10	
G[2]: 71.480003	L[1]: 94.199997	L[6]: 9.000000
	L[2]: 47.799999	L[7]: 3.500000
	L[3]: 7.600000	L[8]: 6.100000
	L[4]: 11.200000	L[9]: 13.500000
	L[5]: 29.500000	L[10]: 14.900000

Table 5 Generator-2 and load pair in IEEE 14 bus system satisfying Eq. 1

Sr. no	Combination		
	G/L ratio	Generator (s)	Load (s)
1	1.099185	G2	L2 + L3 + L7 + L8
2	1.0007		L3 + L4 + L5 + L7 + L8 + L9

The load generation pair obtained for case-I and case-II with G/L ratio approach are shown in Fig. 7 (Fig. 9).

The proposed algorithm is executed on IEEE 14 bus test system graph model. The source node 'a' contribution in other nodes is calculated as shown in Table 6.

In the second part of the algorithm, nodes for intentional islanding are found based on the ratio of source weight to node weight as shown in Table 7.

The node combinations having ratio lying between 1.0 and 1.1 are shown in Table 8.

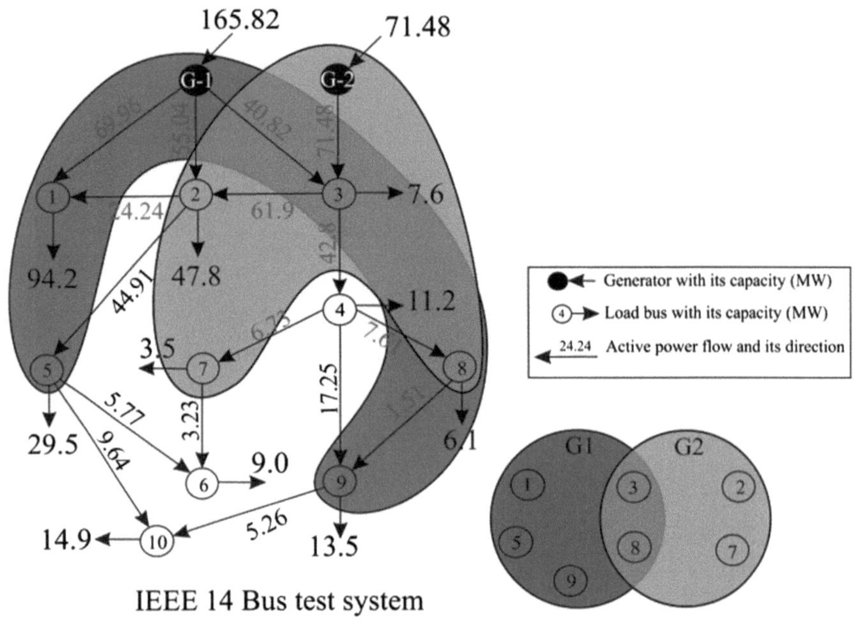

Fig. 9 Generator and load pair in IEEE 14 bus system satisfying Eq. 1

Table 6 Generator-1 contribution in IEEE 14 bus system with ranking

Sr. no.	Node	Load (MW)	Source contribution		
			Node 'a'	%age	Ranking
1	b	7.6	7.04	92.63	5
2	c	47.8	41.86	87.57	8
3	d	94.2	77.85	82.64	10
4	e	29.5	25.81	87.49	9
5	f	3.5	3.26	93.14	3
6	g	11.2	10.45	93.3	1
7	h	6.1	5.68	93.11	4
8	i	13.5	12.59	93.26	2
9	j	9	8.09	89.89	6
10	k	14.9	13.36	89.66	7

Table 7 Generator-1 load pair in IEEE 14 bus system

Source: 1	Node: 10	
Source node 'a'	b: 7.600000	g: 11.200000
	c: 47.799999	h: 6.100000
	d: 94.199997	i: 13.500000
	e: 29.500000	j: 9.000000
	f: 3.500000	k: 14.900000

Table 8 Generator-1 load pair in IEEE 14 bus system satisfying Eq. 1

Sr. no	Combination			
	G/L ratio	Generator	Load (s)	Rank total
1	1.019686	a	b + c + f + g	17
2	1.0007		b + g + e + f + h + i	24

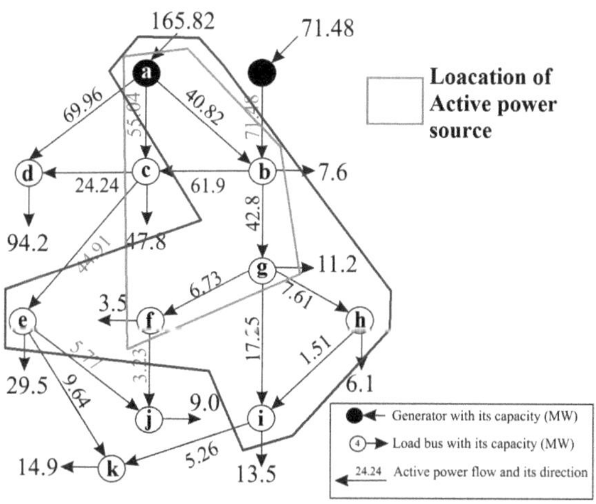

Fig. 10 Generator and load pair in IEEE 14 bus system satisfying Eq. 1 with source location

Based on the node ranking and group of nodes selected from the proposed algorithm, the summation of ranks in first group (b + c + f + g) of nodes is (5 + 8 + 3 + 1 =) 17. Similarly rank total is also found out for second group of nodes. It is observed that the ranking of the nodes: b + c + f + g is higher hence the active source can be located with these nodes when intentionally islanded. The execution time for the second part of the algorithm which depends on real time weight of node was observed for 5 s clock time (Fig. 10).

6 Impact of Islanding on System Stability

The proposed approach is applied on test system. The graph theory is used for finding out generator contribution in each load bus and location for active source based sub-system. The system is tested for three different locations using MATLAB Simulink platform as illustrated in Fig. 11. The location 1 is obtained from proposed algorithm whereas location 2 and 3 are selected arbitrarily.

The voltage response obtained after three phase to ground fault and single phase to ground fault during 1–1.5 s and 2–2.5 s respectively from three locations is shown in Fig. 12. The location of active source is determined using the proposed algorithm.

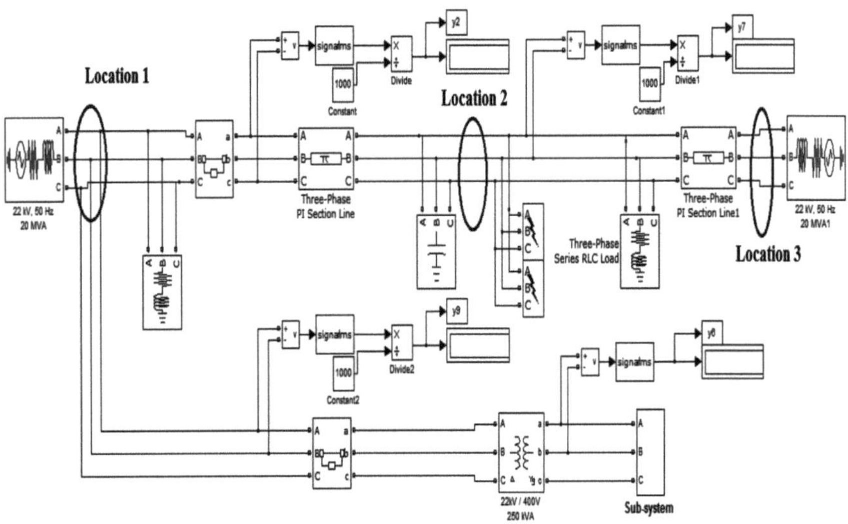

Fig. 11 MATLAB Simulink circuit for testing

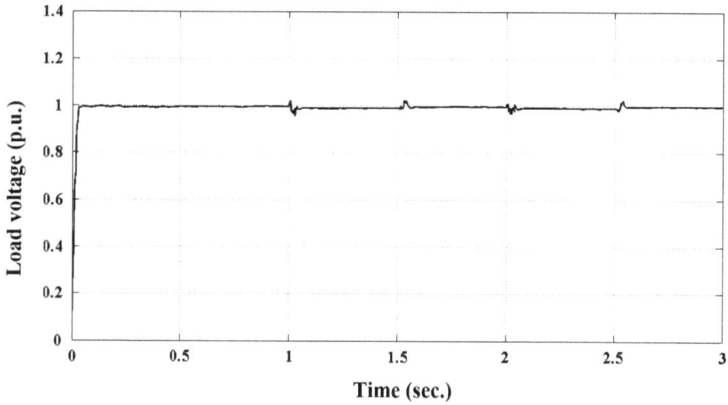

(a) Load voltage at location 1

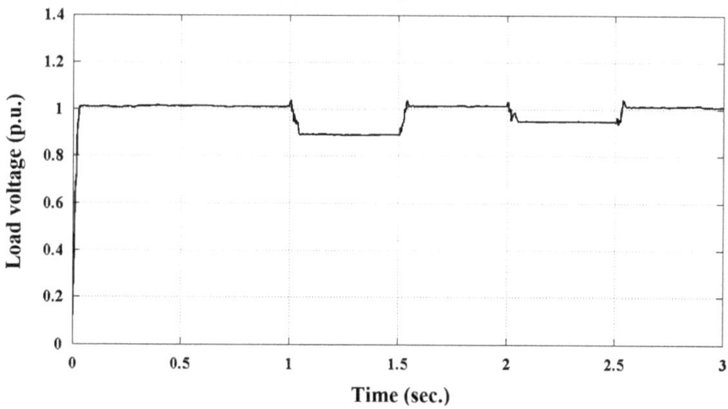

(b) Load voltage at location 2

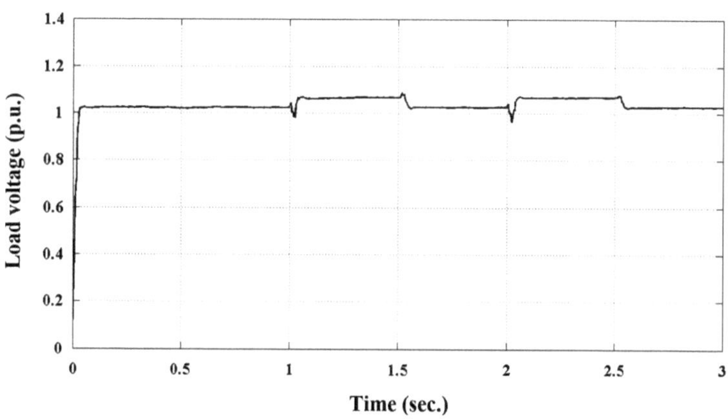

(c) Load voltage at location 3

Fig. 12 Sub system load voltage at three different locations

With reference to Fig. 12a the list of Events is as below:

Sr. no.	Event	Description
1.	$0 \leq t < 1.0$ s	It is a normal working condition of the system. Load voltage is 1.0 pu
2.	$1.0 \leq t < 1.5$ s	Three phase to ground fault is present on the system. The per unit load voltage is decrease slightly during this period after a small spike of voltage due to inductance in the circuit
3.	1.5 s	Three phase to ground fault is clear
4.	$1.5 \leq t < 2.0$ s	It is a normal working condition of the system. Load voltage is 1.0 pu
5.	$2.0 \leq t < 2.5$ s	Single phase to ground fault is present on the system. The per unit load voltage is decrease slightly during this period after a small spike of voltage due to inductance in the circuit
6.	2.5 s	Single phase to ground fault is clear
7.	$2.5 \leq t < 3.0$ s	It is a normal working condition of the system. Load voltage is 1.0 pu

With reference to Fig. 12b the list of Events is as below:

Sr. no.	Event	Description
1.	$0 \leq t < 1.0$ s	It is a normal working condition of the system. Load voltage is 1.0 pu
2.	$1.0 \leq t < 1.5$ s	Three phase to ground fault is present on the system. The per unit load voltage is decrease to 0.9 during this period
3.	1.5 s.	Three phase to ground fault is clear
4.	$1.5 \leq t < 2.0$ s	It is a normal working condition of the system. Load voltage is 1.0 pu
5.	$2.0 \leq t < 2.5$ s	Single phase to ground fault is present on the system. The per unit load voltage is decrease to 0.95 during this period
6.	2.5 s	Single phase to ground fault is clear
7.	$2.5 \leq t < 3.0$ s	It is a normal working condition of the system. Load voltage is 1.0 pu

With reference to Fig. 12c the list of Events is as below:

Sr. no.	Event	Description
1.	$0 \leq t < 1.0$ s	It is a normal working condition of the system. Load voltage is 1.0 pu
2.	$1.0 \leq t < 1.5$ s	Three phase to ground fault is present on the system. The per unit load voltage is increase to 1.05 during this period after a small spike of voltage due to inductance in the circuit
3.	1.5 s	Three phase to ground fault is clear
4.	$1.5 \leq t < 2.0$ s	It is a normal working condition of the system. Load voltage is 1.0 pu
5.	$2.0 \leq t < 2.5$ s	Single phase to ground fault is present on the system. The per unit load voltage is increase to 1.05 during this period after a small spike of voltage due to inductance in the circuit
6.	2.5 s	Single phase to ground fault is clear
7.	$2.5 \leq t < 3.0$ s	It is a normal working condition of the system. Load voltage is 1.0 pu

7 Conclusion

The location of active source is determined using the proposed algorithm. For stable operation of the sub-system during intentional islanding, the active source is located close to the nodes having maximum source contribution with which it is islanded. Hence with the help of an algorithm, shortest path or minimum Relative Electrical Distance for nodes is calculated. When Dijkstra algorithm is used the time taken to calculate shortest path is found to be proportional to the number of nodes in the graph. Consequently it can be seen that proposed algorithm will take less time for a network with large number of nodes.

Reference

1. S.-K. Chai and A. Sekar, "Graph theory application to deregulated power system," Proceedings of the 33rd Southeastern Symposium on System Theory, pp. 117–121, March 2001.

MIX
Papier aus verantwortungsvollen Quellen
Paper from responsible sources
FSC® C105338

If you have any concerns about our products,
you can contact us on
ProductSafety@springernature.com

In case Publisher is established outside the EU,
the EU authorized representative is:
Springer Nature Customer Service Center GmbH
Europaplatz 3, 69115 Heidelberg, Germany

Printed by Libri Plureos GmbH
in Hamburg, Germany